The Mind of Science

The Mind of Science

FROM ARISTOTLE TO EINSTEIN

Michael Sidiropoulos

The Mind of Science

ISBN-10: 1508560692
ISBN-13: 9781508560692
Library of Congress Control Number: 2015903463
Createspace Independent Publishing Platform, North Charleston, South Carolina

Printed in the United States by CreateSpace

Dedicated to all young people aspiring to the study of science.

Table of Contents

Introduction

THIS BOOK AIMS toward an understanding of science with a philosophical perspective that complements the scientific theories. I hope that it will encourage the reader into an appreciation of the philosophical aspects of what constitutes scientific knowledge. The inclusion of mathematical formulas has been avoided except in the discussion of Maxwell's equations. The reason for the exception is that Maxwell's equations, in my opinion, offer the most complete relationship between mathematical formalism and scientific truth. Maxwell's mathematical formalism is presented along with a physical interpretation that explains the mathematics. That discussion does not require anything beyond high-school mathematics, but it can also be skipped altogether without loss of continuity.

The discussions of philosophical ideas in this book do not represent comprehensive reviews of each philosopher's theory. For example, there is nothing about aesthetic or ethical theories. The goal is to relate scientific discoveries with theories of knowledge, and only relevant philosophical ideas are included. It is hoped that our discussion will lead to a better appreciation of the similarities, differences, and complementarity between science and philosophy.

The study of science and philosophy makes us more knowledgeable and powerful and has a positive effect on our morale and our morality. Every generation has a complaint about the new generation—that it is in moral decline. And yet, each new generation marches forward with new, incredible achievements in all areas of human endeavor.

The need for scientific knowledge springs not only from a natural human curiosity to understand our world but also from a genuine desire to help humanity. The great scientific discoveries have given us incredible knowledge about the natural world, have improved our lives through new technologies, and have taught us a great deal about both the capabilities and the limitations of our human perspective.

An Earthly Planet

Birth of the Scientific Method

THE CALCULATION OF the size of the earth is a remarkable scientific achievement, considering when it was made
and the means that were available at that time. A man named
Eratosthenes, chief librarian at the renowned library of Alexandria,
was responsible for this feat. Born in 276 BCE in Cyrene, a Greek
colony that is now part of Libya, Eratosthenes lived in the latter part
of the third century BC. He spent most of his life in Alexandria of
Egypt, a city that was at the height of its glory at the time, having
replaced Athens as the center of the Hellenistic world. His position as
chief librarian allowed Eratosthenes to live a life of learning dedicated
to science and culture. His writings dealt with issues of geography,
astronomy, mathematics, music, and poetry.

In one of his travels Eratosthenes went to Syene in southern Egypt,
now a thriving city named Aswan, built on the banks of the Nile
River. It was June 21, the time of the summer solstice. He noticed that
the sun did not cast a shadow at noon and concluded that the sun's
rays fall vertically at that point in time. On the same date and time in
the following year, he was in his home town of Alexandria and used a
gnomon to measure the length of the sun's shadow. A gnomon is the

part of a sundial that casts the shadow. From the length of the shadow and the length of the gnomon, Eratosthenes calculated the angle of the sun's rays at 7.2 degrees from the vertical line.

Eratosthenes knew from old measurements dating back to Pharaonic times that the distance between Alexandria and Syene was 5,000 stadia. The stadion was a unit of distance and had different values in Greece and Egypt. The Egyptian stadion was equivalent to about 158 meters. Eratosthenes knew that the angle of 7.2 degrees corresponds to the arc of the earth's circumference, the portion of the circumference that covers the distance from Alexandria to Syene. That is, 7.2 degrees is 5,000 stadia, and therefore the complete circumference is found to be 250,000 stadia—or, in present units, 39,500 kilometers. We know today that the circumference of the earth is 40,008 kilometers. The calculation made by Eratosthenes is 98.7 percent of the correct value. The accuracy of his measurement is simply amazing! This is a monumental event in the early history of science.

There are three sources of error in the Eratosthenes calculation that are responsible for the difference of 1.3 percent from the correct value. He assumed that Alexandria and Syene lie on the same meridian—that is, Syene is vertically south of Alexandria. He used the Pharaonic estimates of five thousand stadia for the distance between the two cities, an estimate that was not accurate. He also assumed that the earth is a perfect sphere. We know today that the earth is an oblate spheroid, a sphere flattened on the poles and bulged around the equator. In spite of these errors, the accuracy of Eratosthenes's calculation is remarkable.

Eighteen centuries later, Christopher Columbus, who was familiar with Eratosthenes's work, believed that the earth's circumference was only half of what Eratosthenes had calculated. This misconception is the reason why Columbus thought that he had already circled half of the earth and landed in India. Had Columbus shown a better appreciation for the calculation of Eratosthenes, he would have known that he had discovered a whole new continent.

Eratosthenes also calculated to high accuracy the earth's tilt, which is the angle between the earth's axis and the plane of the earth's orbit around the sun. Another remarkable finding was that the year has 365 days with one extra day every four years. Eratosthenes produced some interesting work in mathematics as well, but his work on astronomy and geography stands out; and he is justly known as the father of geography. Eratosthenes became blind and died of self-induced starvation at the age of eighty-four.

The belief that the earth is spherical is central in Eratosthenes's work. The idea dates back to Greek philosopher Pythagoras, of the famous theorem that bears his name. Pythagoras believed that the gods had chosen the harmonious and perfect shape of the sphere for the world. The reasoning of Pythagoras, of course, is not proof of a spherical shape.

Two hundred years after Pythagoras, the philosopher Aristotle provided a certain kind of scientific evidence. Aristotle noted that those who traveled toward the south would see stars rise up the horizon that were not visible in northern lands. He concluded that this could be explained only if the earth is curved. He also pointed out that during lunar eclipses, the shadow of the earth on the moon has a curved edge.

Aristotle's arguments were convincing; and by the time Eratosthenes did his work one hundred years later, most educated people accepted that the earth is a sphere and not flat. Surprisingly, the flat earth idea gained wide popularity many centuries later in the Middle Ages, to be dispelled only by the travels of Christopher Columbus, which provided irrefutable evidence that the earth is a sphere, as believed by the ancient Greeks.

The work of Eratosthenes was admired throughout the scientific world and became the foundation of several new discoveries in the years following his death. One of the scientists influenced by Eratosthenes was Hipparchus of Nicaea, a Greek mathematician, astronomer, and geographer who was born in Nicaea in 190 BC, just four years after the death of Eratosthenes. Nicaea is the city now known as Iznik in Turkey and lies on the shores of a lake about fifty-five miles southeast of Istanbul.

Hipparchus spent most of his life in the Greek island of Rhodes, where he made most of his astronomical observations. His work provides the earliest known calculation of the distances of the moon and the sun from the earth. Hipparchus calculated the distance from the earth to the moon within an error of 5 percent. He based his calculations on records of a past solar eclipse as seen from two different locations: one in Alexandria and the other at Hellespont. The eclipse seen from Hellespont was a total eclipse, whereas the eclipse seen in Alexandria covered 80 percent of the sun. Hipparchus drew a diagram with triangles showing the geocentric distance from the earth to the moon as well as the distances from Alexandria and Hellespont to the moon and the known distance from Alexandria to Hellespont. He had to invent a new mathematical method to solve these triangles, and that method is known today as trigonometry. Hipparchus is justly regarded as the founder of trigonometry.

Many scholars regard Hipparchus as the father of astronomy as well, as it was his studies and works that first established a scientific method in astronomy. In his work on astronomy, Hipparchus created the first star catalogue. It was a list of the position and brightness of over one thousand stars. At that time the Aristotelian geocentric model of the universe dominated Greek thought. The earth was believed to be the center around which all celestial bodies revolved. Planetary orbits were believed to be perfectly circular. But the model was so rigid that changes in the brightness of the planets, strange retrograde movements of certain planets, and changes in the speed of motion through space could not be explained.

Hipparchus was able to solve some of these problems by studying old Babylonian astronomical observations. The conquests of Alexander the Great had provided Greek scholars with access to records of Babylonian scientists who had done impressive work in mathematics and astronomy. Hipparchus was able to remove many of the contradictions of the Aristotelian model without rejecting the model. He rejected, however, the more correct heliocentric model of Aristarchus of Samos, another great astronomer, who had described a planetary system centered around the sun.

In one of his observations of the night sky, Hipparchus noticed a star he had not seen before at that spot. Stars at that time were not supposed to be created or destroyed, as we know today that they can be. This was a great puzzle that Hipparchus was determined to solve. He went back to a Babylonian star chart created 166 years before he created his own and found that star positions had changed by about two degrees.

Hipparchus developed a theory that stands out in astronomy as one of the finest accomplishments of all time. The theory, which has been extensively verified in modern times, describes a movement of the polar axis, the axis of the earth's revolution about itself. This movement is called precession and is very much like the wobbling movement of the axis of a spinning top. Hipparchus calculated the precession to be thirty-six arcseconds, a little short of the correct value of fifty arcseconds, as we know it to be today. An arcsecond is an arc that has length of one degree divided by 3,600. If we do the math, we will find that one revolution of the polar axis will take about twenty-six thousand years to complete.

We will need to understand this with a visual example. Suppose we are standing at some point on the earth on March 21, the day of the spring equinox. It is just before dawn, and we are looking toward the east. We see a constellation above the horizon at the point where the sun will soon rise. That constellation today is Pisces. About two hundred years from now, the same spot on the same day will be occupied by Aquarius. This shift is called the precession of equinoxes and is due to the circular movement of the polar axis that Hipparchus discovered 2,200 years ago. It is simply amazing! Incidentally, this is what the *Dawning of the Age of Aquarius* is all about, celebrated in songs and Hollywood musicals.

The concept of the universe in classical Greece had harmony, balance, and simplicity, consistent with the aesthetic values of the Greeks. The universe was spherical and had the spherical earth at its center. The stars were fixed in space, but the sun, moon, and planets revolved around the earth. This Aristotelian model is the geocentric universe, meaning that the earth (*gaia*, in Greek) is at the center. The geocentric universe dominated Greek thought, and it could not be otherwise. Even the supreme gods were earthbound, living on Mount Olympus, and no other heavenly body but the earth could be the center of the universe.

A heliocentric universe, one that has the sun (*helios*, in Greek) at its center, had been proposed by another Greek astronomer, Aristarchus of Samos, also known as the "Greek Copernicus." Aristarchus lived in the third century BC and was familiar with Aristotle's universe. It is ironic that the only surviving work attributed to Aristarchus, *On the Sizes and Distances of the Sun and Moon*, is based on a geocentric concept of the universe, and yet Aristarchus is considered the father of heliocentrism. The book is the oldest surviving geometric treatment of celestial observations.

Aristarchus's other book was on the planetary system and had the sun at its center. The book did not survive, and we know of Aristarchus's ideas from later scholars. Archimedes, who was twenty-five years younger than Aristarchus, wrote, "Aristarchus brought out a book consisting of certain hypotheses. His hypotheses are that the fixed stars and the Sun remain unmoved, and that the Earth revolves about the Sun in the circumference of a circle, the Sun lying in the middle of the orbit." Archimedes added that, according to Aristarchus, the universe is many times larger than generally assumed by astronomers, and the fixed stars are at an enormous distance from the sun and its planets.

This heliocentric view of Aristarchus conflicted with the common teaching of his day. Cleanthes, one of the contemporaries of Aristarchus, wrote that "it was the duty of the Greeks to indict Aristarchus of Samos on the charge of impiety for putting in motion the Hearth of the Universe."

The heliocentric universe of Aristarchus gained some following from notable astronomers of the time, but it was Aristotle's geocentric model that would prevail in scientific thought for many centuries to come—for seventeen centuries, in fact, until Copernicus developed his heliocentric universe. Aristotle's universe is described in detail, with some modifications, in the monumental thirteen-volume work *Almagest*, written by Claudius Ptolemy, a Greek-Egyptian mathematician, astronomer, and geographer.

Geocentrism is a flaw in the Greek universe, but there are many elements in that universe that have a striking modernity. The spherical shape of heavenly bodies, motions of the planets, elliptical orbits, and, incredibly, the rotational motion of the earth's polar axis and the movement of the equinoxes are fascinating discoveries, considering the absence of sophisticated measuring instruments and telescopes. And yet these discoveries were all made through experiment and measurement.

This was the first time in recorded history that we see the foundations of the scientific method, the pursuit of knowledge through systematic observation, measurement, and experiment and the formulation, testing, and modification of hypotheses. Aristotle and his disciples had moved away from Plato's pure reasoning into a system where empirical input was the prime source of new knowledge. This is the genesis of empiricism.

For the first time in antiquity, Greek astronomers used scientific measurements and mathematics. The ability to measure celestial triangles was absolutely necessary, and astronomers were fortunate that some great mathematicians, like Pythagoras and Euclid, had laid the foundations of geometry.

Pythagoras is an extremely important figure in the development of mathematics, and yet we have none of his writings, if indeed there were any writings. Some accounts of his life, written centuries after he lived, describe him as a godlike figure with mystic powers. Pythagoras is said to have been the first who described himself as a philosopher, which in Greek means "lover of wisdom."

Pythagoras was born on the island of Samos around 570 BC. This was almost two centuries before Aristotle. Some say that his father was a prominent citizen named Mnesarchos, and others say that Pythagoras was the son of the god Apollo! Well, these are two very different stories, and we can tell right away that any accounts of Pythagoras's life are likely to be partly fictional.

Accounts of his philosophy have been equally conflicting, and anything that we say here or read elsewhere must be thought of as a compromise between widely divergent views of scholars. Like Aristotle, Pythagoras is one of the great polymaths of all time. The English mathematician and philosopher Bertrand Russell has written in a few pages one of the best accounts of Pythagoras and his ideas. Russell writes that "Pythagoras was intellectually one of the most important men who ever lived...one of the most interesting and puzzling men in history."[1]

Scholars generally agree that Pythagoras lived a good part of his life in Croton, a prosperous Greek colony in southern Italy. He had left Samos, his home island, as he disliked the government of Polycrates, a ruthless, corrupt tyrant. At Croton Pythagoras attained extensive influence and followers, who formed a select brotherhood for the purpose of pursuing the religious practices taught by Pythagoras. The brotherhood was a religious group, a philosophical school, and a political association. Similar clubs were soon established elsewhere, and Pythagoreanism became influential in the Greek cities of southern Italy. The Pythagorean order eventually fell out of favor at Croton, and Pythagoras was forced to flee to Metapontion, another Greek city in southern Italy, where he died in his seventies.

1 Bertrand Russell, *The History of Western Philosophy* (New York: Simon & Shuster, 1945), 29.

Religious and scientific views were inseparable for Pythagoreans. They fervently believed in successive reincarnations of the soul into the bodies of humans, animals, and vegetables until the soul became immortal. Pythagoras was the head of the sect, and there was an inner circle of "mathematicians," a select group of followers who lived permanently with the society, had no personal possessions, and were vegetarians. Their basic ideas were as follows: reality is mathematical in nature, philosophy can lead to spiritual purification, and the soul can rise to union with the divine. They also believed that certain symbols have a mystical significance.

Pythagoras believed that all of nature could be reduced to mathematical expressions. He experimented with the monochord, an instrument having a single string, and found that the string produced harmonious tones when the ratios of the string lengths were integer numbers. He played the lyre, a harplike instrument, and used music to help those suffering from illness. Pythagoras could find mathematical symmetries and harmonies in poetry, fiction, and, of course, music. He believed that the number ten was the perfect number as it contained the first four numbers $(1 + 2 + 3 + 4 = 10)$. The *Encyclopedia Britannica* says it very succinctly: "Combining, as it does, a rationalistic theory of number with a mystic numerology and a speculative cosmology with a theory of the deeper, more enigmatic reaches of the soul, Pythagoreanism interweaves rationalism and irrationalism more inseparably than does any other movement in ancient Greek thought."[2]

Driven by a mystical attraction to numbers, the Pythagoreans actually made immense contributions to mathematics. First, we have the famous Pythagorean theorem, which we all had to learn in school. In a right-angled triangle, the square of the hypotenuse is equal to the sum of the squares of the other two sides. It is said that the theorem had been used by Babylonian mathematicians a thousand years before Pythagoras, but the Pythagoreans were the first to formalize it and

2 http://www.britannica.com/EBchecked/topic/485235/Pythagoreanism.

provide a mathematical proof. The square on the hypotenuse was not thought of as a pure number multiplied by itself but rather as a geometrical square shape. The area of this square is equal to the sum of the areas of the two squares constructed from the other two sides. You can reassemble the larger square by cutting up the other two squares.

Also, the sum of the angles of any triangle is equal to two right angles. Furthermore, the sum of the interior angles of a polygon with n sides is equal to $2n - 4$ right angles. The sum of the exterior angles is equal to four right angles, regardless of how many sides the polygon has. In astronomy, the Pythagoreans believed that the earth is a sphere at the center of the Universe and were the first to realize that Venus, as the star of dusk, was the same planet as the star of dawn.

The Pythagorean mix of philosophical rationalism and irrationalism probably sprang out of the rationalism and irrationalism of numbers. If we apply the Pythagorean theorem to an isosceles triangle with sides equal to one unit, the hypotenuse has length equal to the square root of 2. That is not a very good number. It cannot be expressed as a ratio of two whole numbers, so it's irrational, in contrast to Pythagoras's monochord, which produced harmonious tones only when the ratios of the string lengths were whole numbers. The Pythagoreans' quest for mathematical beauty and harmony in the natural world that matched the harmony of numbers is a fascinating idea that marked all of science, from the early astronomical discoveries to Newtonian physics, relativity, quantum mechanics, and eventually the unified theory and the standard model, as we shall see later.

Almost three hundred years after Pythagoras's death, the Pythagorean theorem was generalized to arbitrary triangles by one of the great mathematicians of all time, Euclid of Alexandria. The year of Euclid's birth is not known, but there are some accounts that he taught in Alexandria around 300 BC, a few years after the deaths of Alexander and Aristotle. Euclid's most important work is the *Elements*, which is a compilation of works by earlier scholars, with Euclid's system of rigorous mathematical proofs that remains the standard source of geometric reasoning to this day, twenty-three centuries later. Euclid's system is now commonly referred to as "Euclidean geometry" to distinguish it from other non-Euclidean geometries developed in the nineteenth century. It is said that the *Elements* is the most translated, published, and studied book after the Bible. The thirteen volumes of the *Elements* have been collectively referred to as the most successful and influential textbook ever written.

The *Elements* consist of definitions, postulates, propositions, and proofs, and they cover geometry, number theory, and geometric algebra. The early editions of the *Elements* have raised some criticism as they include proofs that invoke principles that are not listed as axioms or proven propositions. Later editors have included Euclid's implied axioms in the formal axiomatic set. In Euclid's defense, twentieth-century mathematician and historian W. W. Rouse Ball responded to the criticism by saying that "the propositions in Euclid are arranged so as to form a chain of geometrical reasoning, proceeding from certain almost obvious assumptions by easy steps to results of considerable complexity. The demonstrations are rigorous, often elegant, and not too difficult for a beginner. Lastly, nearly all the elementary metrical (as opposed to the graphical) properties of space are investigated, while the fact that for two thousand years the *Elements* was the usual textbook on the subject raises a strong presumption that it is not unsuitable for that purpose."[3] Think of the last phrase in Ball's statement. Isn't this a superb display of British understatement?

Euclid's *Elements* is a true masterpiece in the application of axiomatic logic to mathematics. Its simplicity and beauty is in total harmony with the view of a world of structure, perfection, and certainty. Greek mathematics had a lot to do with aesthetics. The golden ratio is a good example. Two numbers are in the golden ratio if their ratio is the same as the ratio of their sum to the larger of the two numbers.

3 W. W. Rouse Ball, *A Short Account of the History of Mathematics* (New York: Project Dover Publications, 2010), 45.

The mathematical expression is $\frac{a+b}{a} = \frac{a}{b}$. This ratio turns out to be equal to the irrational number 1.618033…The golden ratio was not discovered by Euclid, but it was Euclid who gave its first definition.

It is said that the statues of the Parthenon, made by the sculptor Phidias almost 150 years before Euclid, embodied the golden ratio. The famous series devised by the thirteenth-century Italian mathematician Fibonacci embodies the golden ratio as well. The Fibonacci series is created by adding the last two numbers of the series to produce the next number. The first two numbers are 0 and 1. The next few numbers are 1, 2, 3, 5, 8, 13, 21, 34…and so on. The ratio of any two sequential numbers asymptotically approaches the golden ratio. Later scholars became fascinated by the ratio and found it in the works of Swiss architect Le Corbusier, painter Salvador Dali, and composers Bella Bartok and Erik Satie.

It is known that Abraham Lincoln kept a copy of Euclid in his saddlebag and studied it late at night by lamplight. According to his own account, he said to himself, "You never can make a lawyer if you do not understand what demonstrate means; and I left my situation in Springfield, went home to my father's house, and stayed there till I could give any proposition in the six books of Euclid at sight."[4]

Einstein referred to the *Elements* as the holy little geometry book and recalled that as a boy, he had received two gifts that had a tremendous influence on him: a magnetic compass and a copy of Euclid's *Elements*. Einstein is one of many great people who were tremendously influenced by the *Elements*, and so were scientists like Copernicus, Kepler, Galileo, and Newton as well as philosophers like Spinoza and Bertrand Russell. Many sections of Newton's *Principia* read very much like Euclidean geometry. The logical axiomatic approach and rigorous proofs have made Euclid's *Elements* the cornerstone of mathematics.

---

4 Henry Ketcham, *The Life of Abraham Lincoln*, Project Gutenberg, http://www.gutenberg.org/ebooks/6811

The remarkable achievements of Greek philosophers and scientists laid the foundations of our Western civilization. What was the driving force that motivated those philosophers and scientists to do all this work? It was not glory or money or the need for practical use. It was good old scientific curiosity, a human desire to understand our home planet and our universe.

Stars So Beautiful Bright

Copernicus, Brahe, Galileo, Kepler, and Newton

IT WOULD TAKE seventeen centuries and the brilliant mind of Nicolaus Copernicus to upset Aristotle's and Ptolemy's geocentric universe. Born into a prosperous family, Copernicus lived most of his life in Warmia, Prussia, now part of Poland. He learned several languages and wrote most of his works in Latin, the language of science in his time. Copernicus was a true polymath. He obtained degrees in church law and worked as physician, classics scholar, governor, diplomat, and economist. In economics he developed the quantity theory of money, which became the philosophical basis of Milton Friedman's work on monetarism in the twentieth century. At the age of eighteen, Copernicus enrolled in the University of Krakow, where he became a pupil of Polish mathematician, astronomer, and Aristotelian scholar Albert Brudzewski. The publication in 1543 of Copernicus's work *Revolutions of the Heavenly Spheres* is the beginning of two centuries of profound transformations in mathematics, astronomy, physics, biology, and chemistry. The work of Copernicus marks the beginning of the scientific revolution.

Copernicus's starting point was just an intuition that it does not seem natural for thousands of stars to be spinning rapidly around the earth every twenty-four hours. The faraway stars would have to be moving at impossible speeds. It is far more likely that the stars are actually stationary and the earth itself is spinning. Copernicus then developed the concept that the earth and all the other planets orbit around the sun. The sun is the center of the universe, and what appears as motion of the sun is actually due to the orbital motion of the earth around the sun. The other planets have a similar orbital motion around the sun. Copernicus developed his theory of the heliocentric universe in 1514 but did not publish it until 1543, the year of his death, fearing condemnation of the church. When he did publish it, he dedicated the book to the Pope in an obvious effort to escape condemnation. Though revolutionary for Copernicus's time, when church doctrine and scientific beliefs in academia were dominated by Aristotelian ideas, the new concept was not very different from what Aristarchus of Samos had proposed eighteen centuries earlier.

The great English mathematician and philosopher Bertrand Russell wrote in 1945 that Copernicus does not seem to have known of Aristarchus's heliocentric theory. There is evidence, however, that Russell is incorrect. The name of Aristarchus does not appear in Copernicus's published work, but it appears several times in the original manuscript, which was discovered three hundred years after his death. Apparently Copernicus removed two pages before sending the manuscript for publication. The two pages contained references and credits to the heliocentric system of Aristarchus. There is some debate about this among scholars, but the reason for the removal of the credits by Copernicus is still a mystery. What is important in the work of Copernicus is that man and the earth are dethroned from their cosmological supremacy. Equally important is the fact that the new cosmological paradigm is deeply rooted in ancient Greek astronomy. The contradictions in Ptolemy's version of the Aristotelian universe are resolved with the restoration and refinement of the Aristarchean universe.

From a scientific point of view, this is a smooth evolution of the cosmological paradigm. An evolution that almost follows in text-book fashion the Hegelian dialectic, thesis (Aristarchus), antithesis (Aristotle), and synthesis (Copernicus). From a political point of view, however, it is a bold revolution against a cosmological concept espoused by Christian theology, where man was created by God and placed on the earth, the center of the universe.

The new cosmology of Copernicus did not receive immediate recognition. The fear of religious persecution was partly responsible, especially in Catholic countries, where the Inquisition had virtually ended all scientific discovery. But even in Lutheran countries, acceptance of heliocentrism was met with obstacles. Both Luther and Calvin attacked the new theory as being in disagreement with the Scriptures.

Tycho Brahe was a Danish nobleman with a passion for astronomy. Born in 1546, just three years after Copernicus's death, Tycho was raised by his wealthy, childless uncle and began his university studies at the very young age of thirteen, studying law, philosophy, and various other subjects at the University of Copenhagen. In 1560 Tycho was quite impressed by the fact that a partial solar eclipse had been predicted and happened right on schedule. The predictability of a celestial event against Tycho's own uncertain life had an impact on young Tycho and inspired him to dedicate his life to the study of astronomy.

A flamboyant, hot-tempered young man, Tycho lost a good chunk of his nose during a duel with a fellow student over who was the best mathematician. For the rest of his life, Tycho used a metal nose to cover the disfigurement. His prosthetic nose was made of copper, but he had silver and gold noses to wear on special occasions.

Tycho's uncle died of pneumonia after saving the king of Denmark from drowning. The king showed his gratitude by giving Tycho an entire island and all the resources needed to build an observatory. Tycho married a peasant's daughter in 1573, and the marriage between the nobleman and a commoner scandalized many of his contemporaries.

Tycho lived an excessive, colorful life and died during a banquet under mysterious circumstances in 1601 at the age of fifty-four. There was a theory that he was poisoned because he had an illicit affair with the Queen of Denmark, an affair that may have inspired *Hamlet*, the Shakespearean masterpiece of intrigue, infidelity, and murder in the court of the kingdom of Denmark. Another theory said that Kepler murdered Tycho to get the extensive records of Tycho's astronomical observations. These theories of Tycho's murder have since been discredited, but they are characteristic of Tycho's larger-than-life persona.

Tycho was a master of precision who was never satisfied with the accuracy of astronomical tables available at the time. He built his own instruments for his painstaking observations and measurements, and these instruments became the best devices available before the invention of the telescope. Tycho was able to measure celestial distances to a precision of one arcminute, which is one sixtieth of one degree. It is believed that Kepler's construction of his model of the solar system could not have been possible without Tycho's astronomical records.

In 1572 Tycho observed a bright star that appeared suddenly in the constellation of Cassiopeia, the familiar star pattern that resembles a sitting woman, the queen of the night sky. Some observers believed that the new star was a certain unexplained phenomenon in the area between the moon and the earth, as the world of the stars was considered unchangeable both in the Aristotelian and in the Copernican universe. Tycho had a different opinion; he noticed that the position of the new star in the sky was not changing as it would have had to if it were part of the solar system. Observations over several months convinced Tycho that this was a new star far in the sky, way beyond our planetary system.

Tycho published a book on the new star, which we now know was an exploding star, or supernova. In the preface of his book, Tycho writes, "Oh thick wits. Oh blind watchers of the sky," referring to those who doubted the significance of his discovery for astronomy. Tycho was quite right; this discovery was very significant in the evolution of our knowledge of the universe, as it shattered the dichotomy between the fixed and immutable heavens and the ever-changing planetary system.

Tycho's concept of the universe was different from the Copernican system. In Tycho's system the sun and moon orbited the earth, whereas the planets orbited the sun. Tycho believed that the idea of a moving Earth would be in violation of physics as well as the Scriptures. It is evident that Tycho's system retains features from both Aristotle's geocentric system and the heliocentric system of Copernicus.

For many of Tycho's contemporaries, his system became a viable compromise for keeping attractive features from both the geocentric and the heliocentric systems. Some scholars argue that Tycho's system is mathematically equivalent to Copernicus's system—that is, one can be transformed to the other with a mathematical transformation. Tycho's geocentrism became the leading theory of his day, but its popularity did not last very long. However, the value of his astronomical observations is immense, and his contribution to the scientific revolution of his day is undisputed.

One of the most fervent supporters of the Copernican system was the Italian physicist and astronomer Galileo Galilei. Born twenty years after the death of Copernicus, Galileo was destined to start a new scientific revolution and elevate science to a prominent discipline with concepts and methods of a whole philosophical system. Albert Einstein said that Galileo was the father of modern science. Galileo attended medical school at the University of Pisa but quickly turned to the study of mathematics. While at the university, he experimented with pendulums and falling objects and became interested in astronomy. Galileo challenged Aristotle's assertion that heavy objects fall faster than light ones by dropping balls of various weights from the top of the Leaning Tower of Pisa.

Galileo invented the telescope and used it to make astronomical observations. His observations of the moon revealed that it is not the smooth, perfect sphere envisioned by Aristotle but instead has a rugged surface with mountains and craters. Galileo used his telescope to observe the planet Jupiter. He saw three small, bright points near Jupiter and initially thought they were distant stars, but with subsequent observations, he noticed that they had moved to the other side of Jupiter. He concluded that these were moons orbiting Jupiter. The universe was not geocentric after all. There were celestial bodies orbiting bodies other than the earth.

The discoveries of the moons of Jupiter were initially met with enthusiasm by the top echelons of the Catholic Church, but Galileo's subsequent adoption of the Copernican heliocentric universe took him to a trial before the Inquisition in Rome. Galileo was sentenced to house arrest and was later allowed to retire in his villa near Florence. After 359 years, in 1992, Galileo was finally cleared of heresy and pardoned by Pope John Paul II with a statement expressing the regret of the church for the way Galileo had been treated. Galileo became a martyr of objectivity and is probably the only scientist in history who has achieved the status of a popular hero.

With the invention of the telescope, Galileo revolutionized astronomy. His experiments on motion, falling bodies, and trajectories prepared the ground for the development of classical mechanics by Isaac Newton. Apart from his discoveries and his championing of Copernican cosmology, Galileo's importance lies in the revival of science and the scientific method. Galileo's method has its roots in Aristotle's scientific work, but Galileo laid the foundations of the modern scientific method with his innovative combinations of mathematics and experimentation.

The life of Galileo set the stage for a new cultural movement of intellectuals, a movement that emphasized reason and individualism rather than dogma and tradition. The new movement swept Europe like a huge tidal wave and marked the beginning of the scientific revolution and the beginning of the Age of Enlightenment.

Galileo had many successors, and the most prominent one, Johannes Kepler, was to play an important role in the refinement and widespread acceptance of the Copernican universe. Kepler was born in 1571 in Weil der Stadt, a small town near Stuttgart. He was born prematurely and was weak and sickly as a child but impressed everyone around him with his phenomenal mathematical abilities. His parents struggled to make ends meet, but they managed to foster young Kepler's intellectual interests and to provide him with a good education.

Kepler attended the University of Tubingen, where he studied theology, philosophy, languages, and science. It was at Tubingen where Kepler was introduced to Copernican cosmology. Kepler achieved an immortal position in the history of astronomy with his three laws of planetary motion. The first two were published in 1609 and the third ten years later. The first law states that the orbit of a planet is an ellipse, and the sun occupies one focus of the ellipse. The second law states that the line joining the planet to the sun sweeps out equal areas in equal times. The third law states that the square of the orbital period of the planet is proportional to the cube of its average distance from the sun.

Let us now make a mental detour and think about what many people say today about the relativity theory and quantum physics—that they are not intuitive. We will have much more to say about this later in the book. If we think about the elliptical orbits of planets, we must admit that in Kepler's time, they were counterintuitive as well. This is probably why circular orbits, both in the Aristarchean-Copernican and in the Aristotelian-Ptolemaic universe, stood the test of time for so long. The circle is a perfect shape, and circular orbits satisfied common sense and aesthetic values. We have here an important epistemological issue about the nature of intuition: is intuition an inherent ability of the mind, or is it empirical, shaped over time by new sensations and new experiences? How does intuition affect our ability to acquire new knowledge?

The counterintuitiveness of Kepler's first law would be an impediment to its wide acceptance, but this is a problem with all bold discoveries in science. The second law is equally shocking: the planet does not orbit the sun with a constant velocity. This is not elegant at all; all notions of balance and aesthetic symmetry are shattered. The planet moves faster at the perihelion, the closest distance from the sun on the elliptical orbit, and moves slower at the aphelion, the farthest from the sun on the orbit. At all other points on the orbit, the planet's speed is somewhere between these two extremes. Whereas the first two laws deal with each planet separately, the third law affords a basis of comparison of the movements of all planets. It says that the length of a planet's year and its average distance from the sun have the same mathematical relationship for all planets.

The movements of the planets could now be described in mathematical equations. The value of this accomplishment cannot be overestimated. Mathematicians, physicists, and astronomers could now use the equations to discover new relationships and explain phenomena that could not be explained before. One such person was Isaac Newton, who developed his theories of motion and gravitation and then used these theories to derive Kepler's laws. This was an astonishing confirmation of Kepler's heliocentric cosmology as well as Newton's own theory. Newton went further and used his theory and Kepler's laws to explain the trajectories of comets, the tides, the precession of equinoxes, and other planetary phenomena.

Many scientists believe Newton to be the greatest scientist of all time. Polls conducted among scientists today consistently show Newton and Einstein in the top two positions on the scale of scientific greatness. Newton was born in 1642, one year after Galileo's death, in a small village in Lincolnshire, some thirty miles east of Nottingham. He grew up in the midst of the political turmoil of the Civil War of 1642. He was not a good student at school, but the signs of intellectual brilliance, especially in mathematical aptitude, were quite evident in his younger years.

We cannot resist telling once again the well-known story about Newton and the apple tree. It is a story often dismissed by Newton historians—but a story that probably has some truth in it. We will tell the story now and will augment it with a bit of our own fantasy: While relaxing under an apple tree on the family farm, Newton saw an apple fall from the apple tree. Whether or not the apple hit him on the head, as the story goes, is unimportant. Newton questioned why the apple falls to the ground while the moon remains in its orbit around the earth. Newton was familiar with Galileo's work on projectiles and decided to conduct his own projectile experiments. He picked up an apple and threw it as far as he could to watch the parabolic shape of its trajectory and eventual fall to ground. After repeated throws with ever-increasing force, he concluded that it should be possible to throw an apple, if he could muster the necessary force, in a way that the change in its trajectory exactly matches the change in the earth's curvature.

This is how the theory of gravity was born. The moon is in a free fall, attracted by the earth's gravity. It does not quite make it, as its parabolic displacement during the fall is matched exactly and canceled by the earth's curvature. As the projectile-moon makes its parabolic turn and begins to fall, the earth's curvature changes. The moon is doomed to a perpetual fall and never makes it to ground. Incidentally, this is how modern satellites are placed in orbit around the earth. They are launched with sufficient force and at such an angle that their parabolic fall to the earth has a vertical component that is matched exactly by the vertical component of the earth's curvature.

At the age of nineteen, Newton entered Trinity College at Cambridge University, where the academic curriculum focused mainly on Aristotelian logic, ethics, and physics. He supplemented his studies with readings of Descartes and astronomers like Galileo and Kepler. When the university closed because of the plague that was spreading across Europe, Newton spent two years at home studying problems in mathematics and physics. This is where he developed the first concepts of infinitesimal calculus, gravitation theory, and theory of optics.

When he returned to Cambridge, Newton began work on the *Mathematical Principles of Natural Philosophy*, the book that would change man's understanding of the universe forever. The book was published in 1686, and its importance was appreciated quickly. He was now a celebrated public figure and was knighted by Queen Anne in 1705. Newton spent his final years in London and died there in 1727 at the age of eighty-four.

Newton's laws of motion and the laws of universal gravitation are the foundations of what we refer to as classical mechanics. We also make reference to Newtonian mechanics, which means the same thing. Newton's first law of motion states that an object either remains at rest or continues to move at constant velocity unless acted upon by an external force. The second law of motion states that the sum of all forces acting on an object is equal to the mass of the object multiplied by its acceleration. The third law of motion states that for every action there is an equal and opposite reaction. The law of universal gravitation states that any two objects exert a force of attraction on each other. The direction of the force is along the line joining the two objects. The magnitude of the force is directly proportional to the product of the two masses and inversely proportional to the square of the distance between the two objects.

Newton's laws are a phenomenal scientific achievement, and they are the very foundation of all physics. They rationalized Kepler's planetary system and answered important questions, but they also raised many issues, some of which remain unresolved to this day. How can an object exert a force on a separate object far away? Is there a transmitting medium for this force? How is this force different from the force that keeps matter "glued" together? We will come back to this when we discuss strong and weak nuclear forces later in the book. As astounding as it sounds, physicists still do not fully understand gravity. What is lacking is an understanding of the nature of the force of attraction between masses, how this force is generated, and how it is transmitted to cause impacts.

Rise of Empiricism

Aristotle, Locke, Berkeley, and Hume

THE SCIENTIFIC METHOD and discovery is probably quite a bit older than Aristotle, but the great Greek philosopher provides the first known paradigm in history of what we now call the scientific method. According to the *Encyclopedia Britannica*, Aristotle was the first genuine scientist in history, and every scientist is in his debt.

Born in 384 BC in Stagira, fifty-five miles east of the port city of Thessaloniki, Aristotle is the greatest polymath of all time. His writings span the widest spectrum of human knowledge imaginable—biology, zoology, physics, geography, astronomy, metaphysics, logic, ethics, aesthetics, poetry, theater, and music. Have we left anything out? Oh yes—linguistics, rhetoric, politics, and government.

When Aristotle was a youth, his father, Nicomachus, died, and Aristotle was raised by a guardian until the age of eighteen, at which time he joined Plato's Academy in Athens. The Academy was a school of philosophy founded by Plato, where no doctrines were taught but problems were posed to be studied and solved by the students. The atmosphere at the Academy was unusually liberal by ancient standards. There was no clearly defined academic curriculum and no clear distinction between teachers and students. Can you think of any kids today who wouldn't want to be in a school like Plato's?

The words "academy" and "academic" that are so common in our everyday language have their roots in Plato's Academy, which was named after Academus, a legendary hero in Greek mythology. Plato's academy created the ideal learning environment where a creative mind like Aristotle's could flourish. Italian renaissance painter Raphael's wall painting titled *The School of Athens* is an excellent portrayal of the intellectually vigorous atmosphere at the Academy.

Aristotle's Greek name is "Aristotelis." It is a composite name made up of two words, Aristos (best) and Telis (purpose). So Aristotle means "best purpose." Quite appropriate for a man guided by such great purpose throughout his life! Aristotle stayed at Plato's Academy for about twenty years and became the Academy's intellectual powerhouse. On Plato's death, Aristotle left the Academy and moved back to the palace of King Philip of Macedonia, where Aristotle had spent part of his childhood when his father, Nicomachus, was personal physician to King Amyntas, Philip's father.

Aristotle established himself as the head of the Royal Academy in Macedonia and became the tutor of Philip's thirteen-year-old son Alexander, better known today as Alexander the Great. In this miraculous play of history, destiny brought together the greatest philosopher of all time with the boy who would turn out to be the most admired military genius in history. One year after Alexander became king of Macedonia in 335 BC, Aristotle returned to Athens, where he established his own school, the Lyceum. Following Alexander's death in 323 BC, anti-Macedonian sentiment in Athens forced Aristotle to flee to the island of Euboea, where he died of natural causes one year later. It is estimated that Aristotle wrote about four hundred works, of which only a small fraction has survived. We know of his work mostly from these surviving books and those of his disciples and later scholars.

Aristotelian philosophy has its roots in Plato's ideas, as expected, but Aristotle's development gradually took him on a philosophical path of his own. Plato and Aristotle are the two pillars of Western philosophy, and no student of philosophy today can escape a detailed study of Platonic and Aristotelian ideas. Now, who is the greatest between the two? English philosopher and mathematician Alfred North Whitehead wrote the remarkable dictum that the safest general characterization of the European philosophical tradition is that it consists of a series of footnotes to Plato! On the other hand, Russian American novelist and philosopher Ayn Rand accredited Aristotle as the greatest philosopher in history.

As philosophers of science, we may be inclined to side with Rand, but we do not really need to take sides right now. Our discussion of the philosophy of science will naturally show a stronger affinity with Aristotelian logic than with Platonic ideas. And yet a basic discussion of Plato's idea of forms is absolutely necessary. The most important metaphysical and quasi-epistemological idea in Plato is the theory of forms. It refers to the belief that the world that we see is not the real world but only an image. The forms are abstract representations of all the things that we see around us. The world of forms resides in our minds and is the cause of the apparent world, which constantly changes. The only fixed world is the world of forms in our minds.

Here is where we have our first objection: if Plato had said that the apparent world is not the effect but the cause of the forms, we might be in complete agreement with his idea. A possible interpretation of Plato's idea is that the universe would cease to exist if the human race disappeared. It would certainly cease to exist in the human mind, but no logical person really believes that the universe would physically disappear. Can our objection be proven? Probably not; it is just an intuitive guess!

But our interpretation may be wrong. There are other interpretations that drive the appreciation of Platonic ideas much closer to modern epistemological concepts. In our discussion of American philosopher Charles Peirce, later on, we will see a striking resemblance between Peirce's theory of knowledge and Platonic ideas. That is quite remarkable, considering that Peirce is considered an Aristotelian philosopher.

In other words, there is a certain risk in interpreting Plato. He is not as dogmatic as we might think. His ideas are often propositions for discussion, and they are not necessarily part of a systematic theory. Like his teacher Socrates, Plato promoted the discovery of truth through dialogue. If Plato were present with us today, he might be engaged in our objections, and our discourse might lead both parties to unforeseen conclusions.

If we trace the evolution of logic back in history, we will find its beginning in the works of Aristotle. Logic is the alphabet of philosophical formalism and is at the core of the scientific method. Aristotle was the first known thinker who investigated logic in a methodical, scientific manner; and in doing so, he invented and formalized the rules of reasoning. Aristotle wrote six books that were collectively known as the *Organon*, which means "instrument." These books contain most of Aristotle's work on logic, although his *Metaphysics* also contains ideas and principles of logic. One of Aristotle's most important contributions to logic is the syllogism, which is a logical argument that draws a conclusion from two or more propositions that are known to be true or postulated as being true. For example:

Major Premise: All mathematicians are intelligent.
Minor Premise: Johnny is a mathematician.
Conclusion: Therefore, Johnny is intelligent.

We may restate the above as follows: The entire set A (mathematicians) is a subset of the entire set B (intelligent people). A particular person (Johnny) is a member of A. Therefore, Johnny is also a member of the greater set B. As simple as this argument sounds, it is the basis of deductive reasoning, which means that the conclusion is of no greater generality than the premises. In other words, as the argument moves from the premises to the conclusion, generality becomes less and less. Deductive reasoning is top-down logic, as our argument moves from general to particular. In our example we have two universals (sets with many members): the set of mathematicians and the set of intelligent people. We also have one particular (Johnny), who is a single member and can belong to several universals. Aristotle noted that in a sentence of a syllogism, the universals can be either subject or predicate, while the particular can only be a subject. We will observe that the syllogism has two premises with two terms each. Two of the four terms are common to the premises. The conclusion is a categorical sentence having as terms the two terms that are not common in the premises.

Aristotle developed his deductive logical structure, but he was equally interested in inductive logic, the bottom-up approach, where you discover isolated empirical facts through search and experiment and then develop a generalization or theory that explains all the facts. In his own scientific work, Aristotle's approach was primarily inductive. He cut up insects and fish to look into their internal structure and wrote down his findings, which he categorized according to the main features of each species. This is an inductive approach.

Aristotle was the first major philosopher to argue convincingly for the role of chance in the natural world. First, there are causes that lead to events. There is a causal chain going back to the first cause. This is what we call, in our modern language, "cause-and-effect" relationships. But Aristotle did not subscribe to the simplistic idea that every event has a single cause. Within a chain of events, there are accidents caused by chance.

Aristotle was aware of the boldness of his idea. He noted that earlier scholars had no place for any randomness in their explanations of phenomena. It is quite astonishing that back in the third century BC, we have randomness as a driving force, a concept that had to wait until the nineteenth century, as we shall see in our discussion of American philosopher Charles Sanders Peirce. Randomness is widely considered today as an important factor in modern quantum mechanics. Aristotle's idea of the role of chance is the genesis of indeterminism in philosophy and science.

English philosopher Bertrand Russell, one of the founders of twentieth-century analytic philosophy, has words of approval for certain aspects of Aristotelian philosophy and disapproval for others. In one of his comments on Aristotle's deductive logic,[5] Russell presents the phrase "the present King of France is bald" to show that such a phrase leads to difficulties. The problem for Russell is that France has no king at present, which, as far as we know, is correct! Russell believes that the phrase implies the existence of a king.

It is rather obvious that Russell is correct on the substance. When we say that the present king of France is bald, we definitely convey our belief that France has a king. But here we have a bit of a conflict between structure and substance. The logical structure can be correct even if one or more of the premises are false. In fact, Aristotle wrote in his book *Categories* that an assertion whose subject does not exist must be false.[6] It almost seems as though Aristotle fully anticipated objections like Russell's!

We must say that Russell's criticism on this issue does not address the distinction between form and substance and appears somewhat formalistic. It seems as though Russell introduces false empirical input into a formalistic relation and then concludes that the relation is false.

---

5 Andrew David Irvine, "Bertrand Russell", *The Stanford Encyclopedia of Philosophy* (Spring 2015 Edition), Edward N. Zalta (ed.), URL = http://plato.stanford.edu/archives/spr2015/entries/russell

6 Aristotle, *Categories*, chapter 10, https://ebooks.adelaide.edu.au/a/aristotle/categories/.

The value of a logical structure cannot be refuted by the possibility of misuse.

We saw earlier how Aristotle's geocentric concept of the universe prevailed over the heliocentric concept of Aristarchus, taking seventeen centuries and the genius of Copernicus to restore the order in favor of the heliocentric model. Aristotle's intellectual stature and influence was such that many of his theories were accepted as doctrine without questioning and dominated scientific thought for many centuries. The great German philosopher Immanuel Kant thought that Aristotle had discovered everything there was to know about logic.

Kant was not very far from the truth. Aristotle invented from nothing the entire philosophical area of logic. Aristotelian principles of logic have stood the test of time, and even now, in the twenty-first century, modern courses on logic in our universities are based on Aristotelian principles.

In this book we are focusing on the process of scientific inquiry and discovery, and we will make an effort to find associations between logical structures and specific discoveries drawn from examples of scientific achievements described in the book. We will be learning as we go. Let us take a quick look at the logical structure that led to the discovery of the photon. First, we have Hertz's experiment in 1887, which showed that a spark jumped more readily between two charged spheres if light was shining on them. This photoelectric effect had not been seen before and was not expected to occur based on known theory. So the starting point in our logical structure is not related to some theory; it is a purely empirical fact discovered by experiment: a charged metal sphere close to another charged metal sphere creates a spark when light shines on it. Repeated experiments by others confirmed Hertz's findings. Then along came Albert Einstein, who developed the photon theory to explain the result.

So here we have an inductive method. We have gone from the particular to the universal, from the isolated experimental finding to the theory that explains it. But the theory will not gain acceptance for simply explaining one empirical incident. It must be tested and found to explain other similar or related incidents. Here we start on a deductive course as we walk the path from the universal to the particular. Experiments are performed, Einstein's theory explains new experimental findings, and the theory is finally confirmed and widely accepted. We may risk a generalization and say that inductive reasoning is appropriate in the formulation of a theory while deductive reasoning is more appropriate in the testing and confirmation of the theory.

Albert Einstein said that no idea is conceived in our mind independent of our five senses. In our study of the evolution of the scientific process, we will naturally follow the course of empiricism, the idea that the origin of the creation of new knowledge is not the mind but the senses. Simply put, it means that our minds must receive experiences through our senses before they get to do their own work. The process of scientific inquiry and discovery is an empirical process.

It is quite amazing that the next important empiricist after Aristotle would take almost twenty centuries to appear. His name was John Locke, and he is regarded as one of the most influential thinkers of the Enlightenment. Locke was born in 1632 in Somerset near Bristol, England, to a strict middle-class family of the Puritan faith. After completing his secondary education at the prestigious Westminster school in London, he was admitted at Oxford, where he studied medicine and philosophy and became a lecturer of Greek and rhetoric. After his stay at Oxford, Locke became personal physician to the Earl of Shaftesbury, saving his life when a risky liver operation became necessary. Shaftesbury's political involvement led him to his trial and acquittal for treason, and Locke fled to Holland, fearing a similar fate due to his close association with the earl. After a decade of self-imposed exile, Locke returned to England when one of the most successful revolutions in English history, the Glorious Revolution of 1688, restored political power to men who shared Locke's political views. In the meantime he had started writing his two major works, which he completed in England. They were *Two Treatises of Government* and *Essay on Human Understanding*, the former laying out principles that became the foundations of the American, British, and French Constitutions. The *Essay* is Locke's greatest philosophical achievement and the work that set the foundation of modern empiricism.

The *Essay* is divided into four books, of which the first discusses innate ideas, the second traces the origin of ideas, the third deals with language, and the fourth discusses the limits of understanding. Locke attempts to determine the capability of the human mind and the nature of knowledge. He argues against philosophies of knowledge, like those of Plato and Descartes, which claim that the human mind is equipped with a priori innate ideas and principles that are properties of the mind and have not been delivered by experience.

Locke is responsible for the term "tabula rasa," which means that when we are born, our mind is a blank slate, like a book with blank pages that are gradually filled as we go through life and our experiences are processed into ideas. Locke's writing is great philosophical reading, with clear, understandable language, balanced judgment, absence of doctrine, and lack of presumption. Here is an excerpt that shows a part of Locke's argument against the belief that ideas are innate:[7]

> Had those who would persuade us that there are innate prin-
> ciples not taken them together in gross, but considered sepa-
> rately the parts out of which those propositions are made,
> they would not, perhaps, have been so forward to believe they
> were innate. Since, if the ideas which made up those truths
> were not, it was impossible that the propositions made up of
> them should be innate, or our knowledge of them be born
> with us. For, if the ideas be not innate, there was a time when
> the mind was without those principles; and then they will

---

7 John Locke, *Essay on Human Understanding*, (Project Gutenberg, 2004), Book 1, Ch. 4, URL = http://www.gutenberg.org/cache/epub/10615/pg10615.html

not be innate, but be derived from some other original. For, where the ideas themselves are not, there can be no knowledge, no assent, no mental or verbal propositions about them. If we will attentively consider new-born children, we shall have little reason to think that they bring many ideas into the world with them. For, bating perhaps some faint ideas of hunger, and thirst, and warmth, and some pains, which they may have felt in the womb, there is not the least appearance of any settled ideas at all in them; especially of ideas answering the terms which make up those universal propositions that are esteemed innate principles. One may perceive how, by degrees, afterwards, ideas come into their minds; and that they get no more, nor other, than what experience, and the observation of things that come in their way, furnish them with; which might be enough to satisfy us that they are not original characters stamped on the mind.

One of Locke's basic ideas is that the way we acquire any knowledge is sufficient proof that the knowledge is not innate. Locke refutes the argument that common ideas are innate by suggesting that they are not shared by infants, children, and idiots and are, therefore, not common. No ideas are naturally imprinted on the mind, and there are actually no principles or ideas that are accepted by every human being. Locke proposes that the most basic units of knowledge are simple ideas that are acquired exclusively from experience and combine in three different ways to form more complex ideas: comparison, combination, and abstraction. Our mind forms ideas about the world only through impressions that enter it through our five senses. The mind has the ability to remember past impressions and compare new and old impressions, to make judgments, to refine by abstraction complex ideas into simpler ideas, and to enlarge a simple idea into a complex one by repetitive impressions or by discovery of new impressions.

In Plato's philosophy, all objects, including humans, animals, mountains, and trees, are mere shadows of forms that preexist in our minds. The tree is a particular object, but its treelike qualities are aspects of the ideal form. In Locke's philosophy, the treelike qualities are the result of an abstraction. Once we have observed more than one tree, we begin to distinguish their common features from their particular characteristics, and by abstraction the common characteristics become the form of the tree. The difference with Plato, evidently, is that in Plato the form preexists in the mind and is the reality of the world, whereas in Locke it is the opposite. The particular object is the reality, and the form is a mental abstraction developed from experience. In the third book of his *Essay*, Locke points out different types of language abuse that obscure the communication of ideas. One of the most common abuses is the use of different definitions of words by different people, leading to misunderstanding and superficial disagreement. Locke pays particular attention to the need for communicative language with the use of clear definitions common to all.

In the fourth book, Locke discusses problems of knowledge but also gets into ontological issues, what we normally refer to as "metaphysics." Does the world exist outside of our minds? To many of us, this is quite a naïve question, and yet it has occupied many a great philosopher through history. We know that the universe existed for billions and billions of years before the human race came into existence, and there is plenty of evidence that the world continues to exist as our minds die when we die, but we also know that our beloved philosophers are not the type of folks who are easily impressed with simple ideas! In all fairness to Locke, he seems to hold contempt for metaphysics rather than embrace it.

We have seen that Bertrand Russell has a word to say about everyone and everything. Sometimes we agree and sometimes not, but we always find his thoughts interesting. His review of Locke's philosophy is quite sympathetic on some issues—perhaps not surprisingly so, as Russell is also one of the great philosophers along the same line of empiricism drawn by Aristotle and Locke. We will come to Russell's own philosophical contributions later on, but in the meantime, we will enjoy and be enlightened by his commentary. We have many philosophers in the twentieth century, but few present a twentieth-century view on classical philosophy as eloquently as Russell does. He writes:[8]

> One of the great historic controversies in philosophy is the controversy between the two schools called respectively "empiricists" and "rationalists." The empiricists, who are best represented by the British philosophers, Locke, Berkeley, and Hume, maintained that all our knowledge is derived from experience; the rationalists, who are represented by the Continental philosophers of the seventeenth century, especially Descartes and Leibniz, maintained that, in addition to what we know by experience, there are certain "innate ideas" and "innate principles," which we know independently of experience. It has now become possible to decide with some confidence as to the truth or falsehood of these opposing schools. It must be admitted, for the reasons already stated, that logical principles are known to us, and cannot be themselves proved by experience, since all proof presupposes them. In this, therefore, which was the most important point of the controversy, the rationalists were in the right. On the other hand, even that part of our knowledge which is logically independent of experience (in the sense that experience cannot prove it) is yet elicited and caused by experience. It is on occasion of particular experiences that we become aware of the general laws which their connections exemplify. It would certainly be absurd to suppose that there are innate principles in the sense that babies are born with a knowledge of everything

8 Bertrand Russell, *The Problems of Philosophy*, Project Gutenberg website, URL = http://www.gutenberg.org/files/5827/5827-h/5827-h.htm

which men know and which cannot be deduced from what is experienced. For this reason, the word "innate" would not now be employed to describe our knowledge of logical principles. The phrase "a priori" is less objectionable, and is more usual in modern writers. Thus, while admitting that all knowledge is elicited and caused by experience, we shall nevertheless hold that some knowledge is "a priori," in the sense that the experience which makes us think of it does not suffice to prove it, but merely so directs our attention that we see its truth without requiring any proof from experience. There is another point of great importance, in which the empiricists were in the right as against the rationalists. Nothing can be known to exist except by the help of experience.

Russell makes two statements that appear contradictory. First he states, "It must be admitted, for the reasons already stated, that logical principles are known to us, and cannot be themselves proved by experience, since all proof presupposes them." A few lines later, Russell says, "Nothing can be known to exist except by the help of experience." Is Russell engaged in a circular, fallacious argument? Are logical principles among those innate or a priori ideas that are already imprinted in our mind as we are born? Does the infant have reasoning ability as a result of the possession of logical principles? Do these logical principles remain fixed as the child grows up, or are they expanded in depth and scope through experience? These are some of the questions raised by Russell's statements.

Locke's influence on his contemporaries and on later philosophers was enormous. We have already mentioned the influence of his political theories on the shaping of the American, British, and French Constitutions. His philosophy of knowledge influenced Hume, Berkeley, and Kant, among others, and became the theoretical foundation for modern empiricism, pragmatism, and logical empiricism.

George Berkeley is a major philosopher, even though his basic premise poses quite a challenge to our intuition and common sense. Berkeley held the view that material objects do not exist unless they are perceived. It almost sounds nonsensical. How can something be perceived unless it exists? However, Berkeley's argumentation is so ingenious, it almost seems irrefutable. Berkeley is always grouped together with British empiricists Locke and Hume. The reason is simple: perception (or sensation) is the source of all reality.

Berkeley was born in 1685 near Kilkenny, Ireland. He completed his secondary education at Kilkenny College and entered Trinity College in Dublin, graduating in 1704. He remained at Trinity until 1724 as a tutor and Greek lecturer, and this is where he wrote his major philosophical works. In 1710 Berkeley published his masterpiece *Principles of Human Knowledge*, a work that is still regarded as the best argued expression of metaphysical idealism. Three years later he published the *Three Dialogues between Hylas and Philonous*, wherein he makes a deliberate effort to defend the ideas developed in the *Principles*. Berkeley earned a doctorate in divinity in 1721 and was appointed bishop in Dublin in 1734.

There is a poetic streak in Berkeley. Let us read an excerpt from his opening of the first dialogue in his *Three Dialogues*:

Can there be a pleasanter time of the day, or a more delightful season of the year? That purple sky, those wild but sweet notes of birds, the fragrant bloom upon the trees and flowers, the gentle influence of the rising sun, these and a thousand nameless beauties of nature inspire the soul with secret transports; its faculties too being at this time fresh and lively, are fit for those meditations, which the solitude of a garden and tranquillity of the morning naturally dispose us to. But I am afraid I interrupt your thoughts: for you seemed very intent on something.

Do you think that Byron and Goethe might be envious of George Berkeley, the poet? The dialogue is between Hylas and Philonous. In Greek, Hylas means "matter." Philonous means "friend of reason." Philonous is supposed to be an impersonation of Berkeley himself. Hylas is believed to be an impersonation of John Locke, Berkeley's philosophical opponent.

In this same dialogue, Philonous argues that matter is only known to us by its qualities as we sense them, and it is impossible to imagine matter without these qualities. In the absence of the qualities that we sense, matter loses its essential nature. Berkeley does not suggest that the material world ceases to exist if it is not perceived by humans. He believes that the material world is always perceived by God. We cannot therefore adopt the view that Berkeley rejects the independent existence of the natural world.

Berkeley's arguments seem irrefutable, but only because they have a quasi-axiomatic arbitrariness. Let us consider two hypothetical statements: the claims "God exists" and "God does not exist" are both irrefutable. It is not possible to devise a logical argument to disprove either. At the same time, they cannot both be true, unless the two claims are made from two different axiomatic reference frames having different definitions of "God" and "exists." Therefore, we may say that irrefutability is not sufficient for truth, and we cannot consider Berkeley's conclusions as truths solely based on their irrefutability.

We argue at a later stage that there are two realities. First, there is an absolute universal reality that is independent of human existence. Second, there is a human-perspective reality, as humans are unable to transcend their human perspective in order to achieve awareness of the absolute universal reality. Any human knowledge of a reality will necessarily carry the human perspective. By definition, that is not knowledge of universal reality. Humans can achieve only a human-perspective reality. The "thing to be known" has characteristics that are independent of human existence and are part of universal reality.

Humans perceive certain characteristics of the object that are shaped by their sensory abilities and the processing abilities of the mind. These characteristics may be the same or different from the object's universal characteristics, and it is not possible for humans to know the difference. Any meaningful discussion of reality must be limited to the human-perspective reality, which is the portion of universal reality that the human mind can know; and this portion may be qualitatively different from the ontological reality. There is a difference between what one knows and what is to be known. We have, therefore, two types of reality: universal reality and human-perspective reality.

The above view is very similar if not identical with Berkeley's basic premise. In Berkeley, only God is aware of what we call universal reality, whereas the human mind perceives the human-perspective reality, which does not exist without the workings of the mind. At first glance, Berkeley's views appear nonintuitive. But a careful reading reveals a coherent and systematic philosophy of mind and matter. Berkeley's idealistic empiricism may not have a wide following today, but the reading of Berkeley is pure philosophical delight and is as popular as ever among contemporary philosophers.

Scottish philosopher David Hume is widely regarded as the greatest among empiricist philosophers and was strongly influenced by John Locke, who preceded him by almost a century. Hume was born in 1711 near Edinburgh into a family of moderate means and entered the University of Edinburgh at the unusually young age of twelve. He was not interested in anything except philosophy, literature, and general learning. At the age of twenty-eight, he completed his first major work, *A Treatise of Human Nature*, which turned out to be his intellectual masterpiece and one of the most important books in philosophy. Hume published the *Treatise* in three volumes anonymously, but the work, which advocated a system of morality based on utility, or usefulness, rather than God's word, failed to arouse public interest and debate. When he recovered from his failures, Hume reworked the

Treatise into smaller volumes, believing that style rather than content was the reason for the book's failure. This time Hume was more successful in attracting public interest and was established as an important proponent of a new utilitarian morality. It was not all positive, however, as he was twice denied academic posts in Scotland due to his alleged immorality and atheism.

Hume went to France as assistant to England's ambassador and published *The History of England*, a major work of over one million words that took fifteen years to complete. The book traced events from the invasion of Julius Caesar in 55 BC to the Glorious Revolution of 1688. This time Hume achieved tremendous success and fame, with the book running into six editions. In 1748 Hume wrote *An Enquiry Concerning Human Understanding*, which was a shortened and more readable version of the first volume of the *Treatise*. This is an excellent introduction into Hume's theory of knowledge, and a student of philosophy who wants to read Hume's original works will be well served to start with this book.

Hume has much to say about definitions. Conventional definitions consist of replacing a term with its synonyms, which merely replicates the initial ambiguity and does not reveal the true cognitive content of the term. Hume begins with the term and asks what idea is associated with it. Without such an idea, the term has no cognitive content, regardless of the term's popularity; common use; and prominence in philosophy, theology, or politics. If the term has an associated idea, Hume uses a microscopic method to analyze it, just like a scientist who uses a microscope to analyze matter. Hume looks at breaking down a complex idea to successively simpler ideas and, eventually, to the simplest indivisible idea possible, with the view of associating it with the sensation that produced it. If the process fails, the idea is void of cognitive content. If it succeeds, it will provide a true definition of the term in question. Hume uses this method to show that many of the central concepts of metaphysics lack clearly defined content. Needless to say, Hume does not have much use for metaphysical theories attempting to prove the existence of God, divine creation, the soul, and other similar ideas. We have no reason to believe that any of these ideas are true, as we cannot receive a direct impression of them.

The problem of induction is at the heart of Hume's philosophy of knowledge. Hume writes that induction concerns how things behave when they go "beyond the present testimony of the senses, and the records of our memory."[9] In other words, induction is drawing conclusions from the observed to the unobserved. Hume notes that all such inference rely on the premise that the future will resemble the past.

There are two dimensions to this problem. First, the uniformity of the observation: Hume argues that it is conceivable that nature might stop being regular, and the notion of uniformity cannot be justified. The second dimension of the problem is equally troubling. The principle of uniformity can be proved only by induction. So induction is invoked to prove uniformity, and then uniformity is used to confirm the validity of induction. This is obviously circular reasoning as it invokes in the proof the very idea that we must prove. Hume solves this problem by arguing that inductive inferences are usually made by natural instinct rather than pure reasoning.

---

9 David Hume, *An Enquiry Concerning Human Understanding*, Project Gutenberg website, URL = http://www.gutenberg.org/files/9662/9662-h/9662-h.htm

The Magnetic Charm of Waves

Electrons and Magnets

I N 1842 AN Austrian physicist named Johann Christian Doppler proposed a theory that the frequency of a wave changes when the source of the wave is moving relative to an observer. The frequency is higher during the approach of the source to the observer and lower during the recession. Think for a moment that you are standing on a sidewalk as an ambulance is approaching at high speed. The sound from the ambulance's siren is a high-pitched sound, but it gets even higher as the ambulance gets closer. The sound reaches a screeching pitch just when the ambulance passes in front of you, and then the pitch begins to get lower as the ambulance moves past. This is the Doppler effect, and we have all experienced it but probably did not pay much attention.

Assume that the ambulance is stationary at a distance of 340 meters from you, and its siren is turned on. The first sound wave will reach you exactly one second after its emission from the siren, because we know that sound travels at 340 meters per second. The second sound wave will hit you exactly one second later, and so on. You will hear a new sound wave per second. Now, if the ambulance begins to move toward you, the first sound wave at time zero will reach you in one second. But the second sound wave will reach you in something less than a second as it has to travel a shorter distance now that the ambulance is closer. The frequency that you hear is higher than the frequency of emission. As the ambulance approaches, each sound wave is reaching you at much shorter intervals than one second. The siren produces sound waves at a fixed frequency, but the frequency that you hear is much higher as the siren approaches. The Doppler effect applies to all waves, including light waves emitted by a source that is moving relative to an observer. We shall see later that astronomers use the Doppler effect to determine the velocity of stars by analyzing their frequency spectra.

For its impact on our daily lives, electricity is the greatest discovery of all time. Our society has been so transformed and become so dependent on electricity that the very survival of our large urban populations would be at stake if electricity were to disappear all of a sudden. Aside from its practical applications, the discovery of electricity has led to an extraordinary expansion of our knowledge of the universe.

The first recorded scientific observations of electricity date back to the sixth century BC, when Thales of Miletus, a Greek philosopher, experimented on static electricity produced by rubbing amber, a fossilized tree resin. In fact, the word "electricity" comes from "elektron," the Greek word for "amber." But it was not until the eighteenth century that we saw the first scientific breakthroughs. In 1752 American scientist and scholar Benjamin Franklin conducted his famous experiment with the kite and proved that lightning is a huge electrical spark.

Magnetism had been investigated in the sixteenth century by the English natural philosopher William Gilbert, but it would take another three centuries before a complete theory would be developed to describe magnetism, electricity, and the relation between them.

The man who accomplished this was James Clerk Maxwell, a Scottish mathematician and theoretical physicist. Maxwell is one of the greatest scientists who ever lived. Einstein has said that Maxwell's work is the most profound and the most fruitful that physics has experienced since the time of Newton.

Electricity and magnetism were thought to be two entirely different phenomena until one evening in April 1820 when a remarkable discovery showed a strong interaction between these two fascinating forces. A Danish physicist named Hans Christian Oersted was preparing to give a lecture at the University of Copenhagen. While organizing his lecture materials, Oersted noticed that a magnetic compass was deflected when he switched an electric current from a battery. In addition to being a physicist, Oersted was also a Kantian philosopher, and his mind was fertile ground for processing ideas about the unity of natural phenomena.

Oersted's intuition made him realize that an electric current radiates a magnetic field, but he needed more than intuition—he needed scientific proof. He embarked on a series of experiments and investigations and published his findings, proving that an electric current produces a magnetic field as it flows through a wire. In honor of his contributions, the oersted is the unit by which we measure magnetic induction today. Oersted's discovery had the makings of a true scientific revolution as it created a great deal of excitement and set the stage for more intensive research in the scientific community. Oersted did not provide a mathematical formulation of the relationship between an electric current and the magnetic field it creates, but a French physicist and mathematician named Andre-Marie Ampere was up to the task.

Ampere was a polymath well versed in history, poetry, philosophy, and the natural sciences. He became a professor of philosophy and astronomy at the University of Paris and began experimenting with electricity and magnetism when he heard of Oersted's discovery. Ampere expanded Oersted's experimental work and was able to prove that two parallel wires carrying electric currents attract or repel each other, depending on whether the currents flow in the same or opposite directions, respectively. Ampere was not just a skillful experimenter but also an excellent physicist and mathematician who could generalize the phenomenon into a law of physics and determine the mathematical relations. Ampere's circuital law relates the magnitude and direction of a magnetic field to the magnitude and direction of the electric field that produces it. The law is a cornerstone of electrodynamics and led to Maxwell's complete formulation of electromagnetic theory.

The world of physics was now well on its way to a unification of the two great forces of electricity and magnetism. An English scientist named Michael Faraday conducted a series of experiments that lasted an entire decade, from 1821 to 1831, and discovered that a changing magnetic field induces an electric current in a circuit that encloses the magnetic field. This is known as electromagnetic induction, and it is the discovery of this phenomenon that led to the development of electromagnetic devices such as transformers, generators, and motors.

Ampere's and Faraday's discoveries showed that two different phenomena of nature are actually one and the same phenomenon with two distinct manifestations. Electricity and magnetism exist together, as one causes the other. This was a major scientific breakthrough that led not only to Maxwell 's theory but also to relativity and quantum physics. The discovery of electromagnetism is one of the great scientific revolutions of all time.

James Clerk Maxwell was born to a prosperous family in Edinburgh in 1831. He lost his mother to cancer when he was eight years old, and his father took responsibility for the boy's education. Maxwell was fascinated with mathematics and geometry and wrote his first mathematical paper when he was fourteen. He enrolled in the University of Edinburgh, where he studied logic, metaphysics, mathematics, and natural philosophy. Upon the completion of his studies at Edinburgh, he enrolled in the University of Cambridge. By this time he already had a reputation as a talented mathematician. Maxwell presented a paper to the Royal Society titled *Experiments on Colour*, and before long he was appointed professor of natural philosophy at Marischal College in Aberdeen. He was eventually appointed to a professorship at King's College in London. While at King's College, he investigated issues in optics, astronomy, electricity, and magnetism. He published a textbook titled *Treatise on Electricity and Magnetism* in 1873 wherein he distilled all that was known about electricity and magnetism in a set of four equations.

This set of equations is a true intellectual masterpiece and is probably the most compact, elegant set of equations in all of physics. We are going to take a look at the equations, one by one, and will see that they are not as difficult to understand as they appear on a first viewing. We will also attempt to understand what the physical meaning of each equation is. The laws of nature can be described in ordinary language or in mathematical language. There are pros and cons in each of these methods. Both can give us insights into what scientific truth is.

A mathematical equation is tautological by definition. The left-hand side is the same as the right-hand side. That is the very definition of a tautology. That means that without empirical input, a mathematical expression reveals nothing about nature. In the best scenario, we will have an equation that tells us about quantitative relationships and helps our thinking to compare and discover new relationships between abstract forms. Thankfully, Maxwell's equations are endowed with plenty of empirical input, and his mathematical formalism has a lot to do with nature.

The first equation is $\nabla \cdot E = \rho / \varepsilon_0$. On the right-hand side of the equal-sign, we see the Greek letter ρ, which stands for the electric charge density at a point in space. The denominator ε_0 stands for the permittivity of free space, which is a constant number and is equal to

8.854 x 10^{-12} farads per meter. The permittivity of a material depends on how the material responds to an electric field. Nonconducting materials have much smaller permittivities than do conducting materials. On the left-hand side of the equation, we see the electric field E, shown in bold to indicate a vector. A complete description of a vector always needs two numbers: magnitude and direction. (For example, velocity is a vector.) The inverted triangle with the dot to the left of the electric field E is the divergence operator. It is a mathematical operator, like the multiplication sign, just a bit more interesting. The divergence measures the extent to which the vector E behaves as a source or a sink at a given point. If the divergence is nonzero at some point, then there is a source or a sink at that point. In other words, the only places where the divergence of the electric field is not zero are those locations where charge is present. If positive charge is present, the divergence is positive, meaning that the electric field tends to flow away from that location. If negative charge is present, the divergence is negative, and the field lines tend to flow toward that point. The first equation is also known as Gauss's law. Basically, it tells you that the presence of an electric charge gives rise to an electric field. The law is named after Carl Friedrich Gauss, who formulated it in 1835 but did not publish it until 1867.

The second equation is $\nabla \cdot \mathbf{B} = 0$. This is even more compact than the first equation. Here we have the same divergence operator as before, but here it is acting on vector $\mathbf{B}$, which stands for magnetic field. This equation tells us that the divergence of the magnetic field at any point is zero. The first equation told us that the divergence of the electric field at any point is zero unless there is an electric charge at that point. The second equation is telling us that the divergence of a magnetic field at any point is zero no matter what. We can understand this equation if we remember that all magnets are made up of two poles, north and south. There is no such thing as an isolated magnetic pole. The magnetic field flux lines depart from the north pole and end in the south pole. No concentration of "magnetic charge" is allowed at any point, and none has been found in nature. The amount of incoming field at any point is exactly the same as the amount of outgoing field. The second equation is also known as Gauss's law for magnetism. We can express the first two equations in a slightly different and less rigorous way: the basic entity for electricity is the electric charge, and the basic entity for magnetism is the magnetic dipole.

The third equation is $\nabla \times \boldsymbol{E} = -\partial \boldsymbol{B} / \partial t$. On the left-hand side we have the electric field vector $\boldsymbol{E}$ with our mathematical operator, which is no longer the divergence but rather the curl operator $\nabla \times$. As we might suspect from its name, the curl of a vector is a measure of the field's tendency to circulate around a point. It is almost the opposite of the divergence, which is a measure of the field's tendency to flow away from the point. The right-hand side represents the rate of change of the magnetic field $\boldsymbol{B}$ over time. The meaning of the whole equation is that a circulating electric field is produced by a magnetic field that changes over time. Wherever a changing magnetic field exists, a circulating electric field is induced. Unlike fields that originate from electric charges, induced fields have no origination or termination points; they circulate back on themselves. This equation is also known as Faraday's law of induction. This law was discovered independently by Michael Faraday in 1831 and Joseph Henry in 1832.

Faraday referred to lines of force when he tried to explain electromagnetic induction. Scientists of his day widely rejected his ideas, because they were not formulated mathematically, until Maxwell came along and developed his electromagnetic theory based on ideas from Faraday, Gauss, and others. Faraday's law of induction is captured extremely well in this compact, elegant third Maxwell equation. This ability of changing magnetic fields to induce electric fields is what makes possible the operation of transformers, electric motors, and generators.

The fourth equation is $\nabla \times \boldsymbol{B} = \mu_o (\boldsymbol{J} + \varepsilon_o \, \partial E/\partial t)$. On the left-hand side, we have the curl of the magnetic field **B**. On the right-hand side, we will find what causes this curl—that is, what causes our magnetic field at a point to start rotating about that point. As we see on the right side, it is the sum of the current density $\boldsymbol{J}$ and the term $\partial \boldsymbol{E}/\partial t$ that indicates how the electric field $\boldsymbol{E}$ changes with time. There is a new constant in addition to our familiar constant from the first equation ε_o: the permittivity of free space. The new constant is μ_o, which is the permeability of free space. The magnetic permeability indicates a material's response to an applied magnetic field. In this case we have vacuum as our material, and that's why our permeability constant has the subscript zero.

The meaning of the fourth equation is that a circulating magnetic field is caused by the presence of an electric current and/or an electric field that changes with time. This equation is also known as the Ampere-Maxwell law, named after Andre-Marie Ampere, who had determined in 1820 the relationship between electric currents and magnetic fields. By the time Maxwell began his work in the 1850s, Ampere's law relating a steady electric current to a magnetic field was already well known, but its applicability was limited to static situations involving steady currents. Maxwell extended the applicability to time-varying conditions by adding another source, the term $\partial \boldsymbol{E}/\partial t$ in the equation.

In this discussion of Maxwell's equations, we have used the term "field" without defining it. Everyone is familiar with the term, but the concept is not as clear as it seems, even to physicists. For our purposes we will define it as a region in space and time where a force exists, caused by a mass, an electric charge, a source of magnetism, or a nuclear source. This definition covers all known fields in physics. If any new types of fields are discovered, the definition will need to change. One of the difficulties with the field concept is that a force is created from a source and acts at some distance without any connection to the source and without an apparent need for a transmitting medium. This is still an unresolved mystery in physics.

Maxwell's four equations describe electric and magnetic fields in a most compact, elegant formalism, but they tell us nothing about the effects of fields on electric charges. This gap is filled by the Lorentz force equation. It is not possible to derive the Lorentz force law from Maxwell's equations unless certain strong postulates are stated. We may therefore include it with Maxwell's equations, and then the five equations will provide a complete statement of the classical theory of electromagnetism. The Lorentz force law describes the force experienced by an electric charge in the vicinity of an electromagnetic field. The force on the charge is given by $F = q\,(E + v \times B)$, where q is the electric charge, while E, v, and B are the electric field, velocity, and magnetic field vectors, respectively. We see that the force has two components. One arises from the electric field, the other from the magnetic field.

We have written the four Maxwell equations in their differential form and completed the picture of electromagnetism with the Lorentz force equation. Those who are mathematically inclined may want to take a look at the integral form of Maxwell's equations as well. Their meaning is exactly the same, no matter which form we choose.

It has been well worth our effort to get into the mathematics, which is not very difficult after all. The mathematical formalism becomes easier to understand once we relate it with physical meanings and describe it in plain language. In all their simplicity, compactness, and elegance, Maxwell's equations have linked electricity and magnetism with geometry, topology, and physics and have redefined our perception of space and nature.

The Philosophy of Knowledge

Kant, Hegel, and German Idealism

D AVID HUME'S THEORY of knowledge is the most sophisticated and complete theory of knowledge before Kant, who in
fact was tremendously influenced by Hume and spent much of
his life trying to deal with Hume's conclusions. Kant saw that Hume had
posed a most fundamental challenge to all human knowledge claims.[10]

Immanuel Kant is generally regarded as the greatest philosopher
of the past two thousand years. Born in 1724 in Koenigsburg, Prussia,
a city presently known as Kaliningrad in Russia, Kant was raised in
a Pietist Lutheran family of modest means, in a cultural environment that stressed religious devotion, personal humility, and a strict
interpretation of the Bible. At the age of sixteen Kant enrolled at the
University of Koenigsburg, where he would spend his entire student
and professional life. There was a myth that Kant had never been
farther than ten miles from his birthplace, but there is some evidence

---

10 *Encyclopedia Britannica*, "Skepticism, the 18th Century."

that he actually worked at some time as far as sixty miles away from home!

Kant enrolled as a student of theology but was mainly attracted to mathematics and physics, particularly Newton's physics. When both of his parents died, Kant had to interrupt his studies and become tutor in various families in order to support himself. He was finally able to complete his degree at the age of thirty-one and became lecturer at the university. Kant was so routinely programmatic in his activities that we are told his neighbors set their clocks to the correct time as soon as Kant went out on the street for his daily walk.

Kant became a full professor in 1770. Until then he had written various scientific and philosophical works, gaining a broad reputation in Germany as an important man of science and philosophy. He wrote a theory of earthquakes and essays on wind and geography. His most ambitious scientific writing was on the origins of the solar system. A decade of calm followed when Kant did not write anything. It is believed that he was working his main philosophy during this time while studying the works of other philosophers, and especially those of David Hume.

In 1781 Kant published the *Critique of Pure Reason*, his philosophical masterpiece, one of the most influential works in all philosophy. Kant's work belongs in the Enlightenment, a cultural movement in Europe that was gradually replacing traditional beliefs of religion, superstition, and monarchy by beliefs founded upon science and reason. Kant's words capture the intellectual and social spirit of the Enlightenment: "Our age is the genuine age of criticism, to which everything must submit. Religion and law-giving have often tried to exempt themselves from it, one as too holy to be critically examined, the other as too majestic. But this has made them suspect, and deprived them of any claim to the sincere respect that reason grants only to things that have survived free and public examination."[11]

---

11 Immanuel Kant, *Critique of Pure Reason*, Project Gutenberg website, URL = http://www.gutenberg.org/files/4280/4280-h/4280-h.htm

It is rather unlikely that a "Kant for Kids" book will ever be written. *The Critique of Pure Reason* is a difficult book of dry prose, complex terminology, and densely packed ideas. Two years after the *Critique*, Kant also published the *Prolegomena to Any Future Metaphysics*, which is about half the length of the *Critique*, which has a much more readable style and is an excellent introduction to his philosophy.

Kant makes a distinction between analytic and synthetic propositions. In an analytic proposition, the concept of the predicate is also part of the concept of the subject, as in "a bachelor is an unmarried man." In a synthetic proposition the subject and predicate provide uniquely different information, as in "Caesar was a great general." Another distinction made by Kant is between a priori and a posteriori knowledge. A priori knowledge consists of universal ideas that we have independent of experience, such as our knowledge of mathematics. A posteriori knowledge consists of ideas gained from experience.

We would typically associate synthetic propositions with a posteriori (empirical) knowledge. For example, the idea that "Caesar was a great general" is a complex idea that cannot be registered a priori in our minds. It is an idea that develops through a careful study of empirically known facts of Caesar's personality and achievements. Its development requires historical analysis and value judgments.

Kant, however, argues that there are a priori synthetic propositions and offers arithmetic and geometry as examples. He argues that arithmetic propositions are always synthetic. For example, the proposition 7 + 5 = 12 is synthetic because the concept of 12 is not contained in the concept of 7 + 5. Kant also believes that the proposition 7 + 5 = 12 is a priori, like all arithmetic, because it is a necessary and universal truth we know independent of experience.[12] As we are capable of synthetic a priori knowledge, pure reason can lead to important truths without the intervention of the senses. This is a

12 Immanuel Kant, *Prolegomena to Any Future Metaphysics*, URL = http://www.earlymoderntexts.com/pdfs/kant1783.pdf

major difference between Kant and Hume, as Hume did not believe that there are synthetic a priori truths.

If we want to play devil's advocate for a moment, we might argue as follows: Let us suppose that we replace the left side of $7 + 5 = 12$ by its equivalent of 12. Then we have $12 = 12$. This is obviously a tautology and cannot contain any knowledge. Therefore, the term $7 + 5$ only tells us about the composition of 12 in this particular instance. That is, the totality of 12 is composed of two parts, 7 items and 5 items. This knowledge of composition is the only nontautological element of $7 + 5 = 12$. The only nondefinitional knowledge provided by the equation is the fact that the totality of 12 is made up of two parts, 7 and 5. This knowledge of the composition of 12 can have no meaning unless the numbers 7 and 5 are related to physical items. The expression $7 + 5 = 12$ without empirical input is a tautology.

Let us now think of the addition of two larger numbers having a totality that is not immediately obvious. The result of the addition of 14,149 and 7,652, for example, can be known in two ways only. The first method is to start from 14,149 and add one unit at a time until we have added all 7,652 units, while keeping the score of the resulting sum. We might say that this rather tedious process is as empirical as a process can be. The second method is to apply the arithmetic operation of addition that we all learned in primary school. Is this method not empirical? If not, it is innate. Then why did we have to learn it? We would suggest that the second method is also empirical. That means that the arithmetic operation of addition was developed by experiment and confirmed by trial and error. None of this is innate, none is obvious, and none is known to the newborn or even to the five-year-old preschooler. Arithmetic cannot be an a priori idea. It is a learned mechanism of describing the relations between quantities of objects. If arithmetic were an a priori idea, then why isn't infinitesimal calculus the same? Or set theory? Kant's belief that all arithmetic is a priori is contentious.

We need to be careful when we read philosophy from original works. Philosophers are notorious for using different definitions for the same terms. Kant, for example, often refers to sensibility, perception, and intuition. These three terms are very closely related in Kant, but it is not certain how they differ from one another. One thing is certain: they all relate very closely to the first sensory experience, before the mind takes an active role. Kant's use of "intuition" is quite different from what we mean by the term in everyday language, and this is partly due to the problems of translation from the German text. Kant uses the word "Anschauung," which means view, opinion, or point of view. Kant refers to the perception formed from a first sensory experience of an object. The word "intuition" is probably not the best translation, but we will use it, as it is used widely in Kantian literature.

What we normally refer to as "perception" in our everyday language is an organized, identifiable mental concept of a sensory experience. The active workings of the mind have shaped the sensory experience into a perception. In Kant, however, perception is mere sensation. We can have no knowledge of an object through mere perception unless the perception is actively processed in the mind. Knowledge is created by the active mind, not by the perception itself. Kant makes a clear distinction between the "noumenon," which is the object itself and the "phenomenon," which is our perception of the object. We shall meet this Kantian dichotomy again later when we explore various interpretations of Heisenberg's uncertainty principle.

Here is an excellent example of how many disagreements result from the use of different definitions. Kant believes that the first sensation of an object is not knowledge. It becomes knowledge when the mind has actively worked on it. This is how Kant chooses to define knowledge. Others believe that the first sensation does constitute knowledge of an object. But if we take one extra step and define grades of knowledge, such as good, better, and excellent knowledge, then we may realize that the disagreements are, to a large extent, definitional and semantic. In other words, the first sensation gives a basic knowledge of the object. The first working of the mind improves that knowledge. Further sensations, such as the results of experimental observations, give rise to further workings of the mind, improving our knowledge of the object, and so on. Knowledge of a material or spiritual object does not have one grade; it has many grades. If we agreed on how the grade of knowledge improves through the various stages of sensation, mental process, new sensations, new mental processes, and so on, we might bridge or even close entirely the gap between Kant and the British empiricists.

Mathematical symbolism, which is the language of science, is precise. The language of philosophy must have a similar precision if we are to avoid the ambiguity that gives rise to different interpretations and semantically based disagreements. Such ambiguities explain why Kant's theory of knowledge has received more than its fair spectrum of interpretation. There is justified criticism of his theory, and there is unjustified criticism that is the result of misinterpretation.

Space, time, and causality are fundamental principles in Kant's theory and are not derived from experience. Space and time are pure intuitions inherent in our senses. Causality is a pure intuition inherent in our understanding. In other words, space and time are a form of perceiving, and causality is a form of knowing. All three concepts are a priori ideas.

Another one of Kant's major philosophical contributions is the idea that things as they appear in the mind are different from things as they are. Kant makes a sharp distinction between the natural world and our mental world.

With regard to Kant's idea that space and time are independent from experience, our skeptical devil's advocate might respond with the notion that they are actually derived from experience. He might say that as soon as the newborn opens her eyes, she views her first object spatially arranged, and the combined sensation of "object in space" is processed by the mind. The concept of space is thus formed at that instant for the first time. We must admit that the advocate's idea is just as aphoristic as Kant's.

Goethe described the reading of Kant as walking into a lighted room. Others were not so kind. In his book *The Work as Will and Representation*, German philosopher Arthur Schopenhauer added an entire appendix titled *Critique of the Kantian Philosophy*, in which he outlined his view of the merits and faults of Kant's philosophy. Unlike Kant, Schopenhauer believed that perception of an object occurs without conceptual thought. But here again, the difference of opinion may be caused by the use of different terminology. In Kant, perception is the first stage of the sensory event and occurs before knowledge is created by the mind's intervention. If Schopenhauer had used the word "knowledge" instead of "perception," he might be in genuine disagreement with Kant.

Schopenhauer also accused Kant of obscure writing, which agrees with our general assessment of Kant's use of vague or undefined terms. In order to see Schopenhauer's criticism in the proper light, we must add that Schopenhauer regarded Plato and Kant as the greatest of all philosophers. It is believed that Schopenhauer moderated his critical view of Kant once he read the later editions of the *Critique of Pure Reason*, which had removed some of the contradictions and inconsistencies.

The main difference between Kant and empiricism is that in Kant, the mind is already armed with concepts and ideas before the newborn infant opens her eyes for the first time. In empiricism, the mind only provides the necessary infrastructure to process sensations into concepts and ideas. Kant's philosophy of knowledge became a major force against the rising tide of British empiricism and set the foundation of German idealism, a philosophy that would dominate continental Europe in the nineteenth century.

This idealism would be taken to new heights by another great thinker, Georg Wilhelm Friedrich Hegel. Born in Stuttgart in 1770, the same year that Beethoven was born, young Hegel entered primary school at the age of three, having been taught the essentials by his mother. A voracious reader, Hegel developed very quickly; and at the age of eighteen, he followed family tradition and entered a Protestant theological school attached to the University of Tubingen.

Hegel was not at ease with the school's theological curriculum and was learning more from his own studies. At school he became friends with future romantic poet Hoelderlin and future idealist philosopher Schelling. This was the time of the French Revolution, a turbulent time in Europe, and the three friends had many opportunities to exchange ideas by engaging in intellectual and political discussions. The three watched and discussed the unfolding of the French Revolution with enthusiasm, and their friendship was a major influence on their philosophical development.

In 1801 Hegel went as an unsalaried lecturer to the University of Jena, near Leipzig. This was a major center of critical philosophy, and for the first time, Hegel became interested in Kantian theory. After an affair with his landlady and the birth of a son, Hegel moved to Bamberg in 1807 to become editor of the local newspaper, while his son and the child's mother stayed behind in Jena.

It was in Bamberg that Hegel wrote and published his philosophical masterpiece, the *Phenomenology of Spirit*. Hegel managed to return to his academic career when he was appointed to a post in Heidelberg in 1816 and, two years later, at the University of Berlin, where he established himself as a superb professor of philosophy. His fame spread, and students would come from various parts of Germany to attend his lectures. Hegel lived and lectured in Berlin until his death in 1831.

A good understanding of Hegel's ideas is difficult and time-consuming. His obscure, cryptic language and terminology create great difficulties in understanding his philosophy from the original works. The difficulties in translation from the original German text make the theories even more obscure. Even Bertrand Russell, the supreme logician and analytic philosopher, claimed that Hegel was the most difficult of all philosophers to understand. There were scholars in Hegel's time who believed that Hegel was a charlatan who purposely used incomprehensible language in order to hide the emptiness of his theories.

That view has been totally discredited as more and more scholars have realized that Hegel's work is a solid philosophical system with important implications for literary and cultural theory, historical analysis, the theory of knowledge, religion, aesthetics, and political philosophy. In this book we will not review the entirety of Hegel's philosophy, but we will search for any insights that will help us in our quest to understand the nature of scientific revolutions.

The *Phenomenology of Spirit* is a study of the historical evolution of consciousness as shaped by appearances, images, and illusions. The first phase of consciousness is sense perception, and the final phase and ultimate goal is absolute knowledge. One of the problems that Hegel saw in Kant's theory is the dualism of sense and mind, the clear distinction between sensory perception and mental process that did not allow for much interaction of the two and led to a detachment of knowledge from reality.

In our review of scientific discoveries, we have seen and will see more of the interaction between phenomena and mental perceptions of the phenomena. New aspects of phenomena interact with existing theory through contradictions and conflicts and lead to improvement of the theory and new conflicts and interactions. The evolution of knowledge is organic change through the conflict of contradictions; it is not a succession of dead ends.

Hegel's system addresses the duality distinction with the dialectic triad, which is arguably Hegel's greatest contribution to philosophy. The dialectic consists of three phases: thesis, antithesis, and synthesis. The thesis is an initial situation in an evolutionary process. Any such situation, even if successful, will generate elements that contradict it. Through the conflict between the original thesis and the contradictory elements, a new situation will emerge, the synthesis, which resolves the conflict by integrating their common truths. The synthesis is not the final truth and becomes the new thesis, which will be subject to new contradictions and will sow the seeds of its own destruction. This is how an evolutionary process unfolds.

We need to be careful and not think that the three stages are distinct from each other. The thesis and antithesis coexist, and their conflict generates synthetic elements, which are also present at all times until they become the dominant situation, the new thesis. It is an organic evolution, not a succession of distinctly identifiable stages. The authoritarian pre-1789 monarchy in France coexisted with opposition and anarchy against its tyrannical rule. These antithetical elements created the ultimate conflict, the French Revolution, which overthrew the ruling aristocracy and set the stage for the synthesis, some sort of democratic rule, which seized power while carrying and creating its own contradictions and conflicts.

Hegel's dialectic is an ingenious scheme that can explain much of history. We can apply the dialectic to the evolution of scientific knowledge. Let us take an example from our review of scientific revolutions. Aristotle's geocentric universe is our starting point—it is the thesis. The Aristotelian system had many strikingly modern elements, like the spherical shapes of heavenly bodies, elliptical orbits, the rotational motion of the earth's polar axis, and the movement of the equinoxes. But the system had its flaws. If the earth is the center of the universe, the distant stars must be rotating around the earth at impossible speeds every twenty-four hours. It is far more likely that the stars are stationary, and the earth is spinning. This hypothesis is the antithesis. Resolution of the conflict is achieved in the synthesis, the Copernican universe, which retains many of the elements of the Aristotelian system but resolves its flaws. We see that the contradiction was internal to the Aristotelian universe. In the Hegelian dialectic, contradictions are inherent in things and ideas, which move from inner contradiction to a higher level of integration.

Hegel seeks to achieve unity between opposites. Whereas Kant seeks to keep reality and knowledge apart, Hegel seeks to achieve unity of the two. But isn't this exactly what the scientist is trying to do—achieve unity between reality and knowledge? Each triad of thesis-antithesis-synthesis achieves a higher degree of unity between knowledge and reality, a better understanding of reality, until perfect unity, the Absolute, is reached. Hegel's theory helps us understand evolutionary processes; and, when applied to the evolution of science, it explains and justifies the driving forces of scientific discovery. Also, there is a strong element of ethical consequence in Hegel's philosophy: if a better unity between knowledge and reality can be achieved, the human effort to understand the world is not meaningless.

Light, Radiation, and Radioactivity

From Democritus to Rutherford

THE DISCOVERY OF electromagnetism and Maxwell's formulation of electromagnetic field theory was a new scientific revolution and opened the door to relativity and quantum physics, two theories that are now one hundred years old but are still at the very front of modern physics. The road from Maxwell to Einstein and Planck has many parallel paths and many exciting discoveries that we need to understand before we venture into quantum physics and relativity. The nineteenth century has multiple scientific revolutions that changed our perception of nature.

Maxwell's work generated a great deal of enthusiasm among scientists who saw a vast field of exciting new research opening up. German scientist Heinrich Hertz was one of them. Born in 1857 in Hamburg, Germany, Hertz came from a wealthy, educated family. His father was a respected lawyer who would later become a senator. Hertz showed a strong interest in science and mathematics at a young age. He studied science and engineering, earning a doctoral degree at the University of

Berlin under the eminent physicist Hermann von Helmholtz. Hertz's most important scientific work was carried out at the Karlsruhe Polytechnic, where he became a professor in 1885.

In one of his experiments, Hertz set up two brass knobs separated by a gap small enough to allow electric sparks to leap from one knob to the other. Each knob was connected to an induction coil. A similar setup without the coil was placed a few meters away. When the induction coil connected to the first loop produced a voltage, a spark jumped across the gap, as expected. But Hertz detected a weaker spark across the gap of the second loop that was a few meters away. He concluded that an electromagnetic wave was generated from the first spark and was transmitted to the second loop. He repeated the experiment using a rotating mirror to find the frequency of the invisible wave; and from the frequency, he determined the wave's velocity, which turned out to be equal to the speed of light!

Hertz conducted many more similar experiments to explore the invisible waves and found that they had many of the properties of light, such as diffraction, refraction, and polarization. In one of those experiments, in 1887, Hertz discovered that a spark jumped more readily between two charged spheres if light was shining on them. This was a "photoelectric effect," an extraordinary discovery that would be better appreciated when Einstein investigated it twenty years later. This was the first time that scientists began to realize that light is a wave. The discovery of electromagnetic radiation by Hertz was the most dramatic prediction of Maxwell's theory and is an important breakthrough that opened the door to major new discoveries. Wireless communications, radio, television, and radar are some of the applications of this discovery. Hertz died of a blood disease at the age of thirty-seven. The unit of frequency is named after him. One hertz is equal to a frequency having one cycle per second.

Visible light is an electromagnetic radiation having frequencies from 430 terahertz (THz) to 790 THz. One terahertz is one trillion hertz (one trillion cycles per second). That's a very high frequency!

Any radiations with frequencies lower than 430 THz, such as radio waves, or higher than 790 THz, such as x-rays, are not matched by the response capabilities of the human eye and are therefore not visible. The red color of light, which has the lowest frequency in the visible spectrum, starts at 400 THz. Violet color, which is the highest end of the visible frequency spectrum, ends at 789 THz. Between these two extremes, we have orange, yellow, green, and blue. Just think, all these colors are different electromagnetic waves traveling at different frequencies but at the same speed! They all arrive together, and our eyes see only their cumulative effect: the whitish-yellowish sunlight we all love.

There are electromagnetic waves of higher frequency able to penetrate solid bodies, like the human body. One familiar example is x-rays. They were discovered in 1895 by German physicist Wilhelm Roentgen, who found them by accident while experimenting with vacuum tubes. Roentgen noticed a glow on a fluorescent screen in his lab when an electron beam in a vacuum tube was turned on. It was not surprising that a fluorescent screen would glow in the presence of electromagnetic radiation, but in this case, Roentgen had the tube surrounded by heavy black cardboard, which was supposed to block the radiation. Roentgen placed various objects between the tube and the screen, and the screen still glowed. When he put his hand in front of the tube, he saw the silhouette of his bones projected onto the fluorescent screen.

Roentgen correctly thought that he had just discovered a new type of radiation, which he called "radiation X" to indicate that it was an unknown type of radiation. This remarkable discovery was the beginning of one of the most important medical advancements in human history. With x-ray technology, doctors could now see through human tissue to examine bones as well as softer tissue such as lungs, blood vessels, and intestines. It is impossible to overestimate the importance of this scientific achievement if we think of all the millions of human lives that have been saved due to the diagnostic capabilities of x-rays.

One year after Roentgen discovered x-rays, a startling new discovery, radioactivity, would rock the scientific community. Radioactivity was discovered in 1896 by French physicist Henri Becquerel, who occupied the chair of physics at the Ecole Polytechnique, a famous engineering school in Paris. Born in 1852 into a family of four generations of scientists, he studied science and engineering and became chief engineer with the Department of Bridges and Highways. Becquerel was attracted to his father's scientific research in the study of various phenomena of light: polarization, fluorescence, and phosphorescence. He had also inherited the minerals and compounds studied by his father.

By 1896 Becquerel was an accomplished and respected physicist and a member of the Academy of Sciences. When he learned of Roentgen's discovery of x-rays, he had his own ready sources of fluorescent materials and began to pursue the study of the new mysterious rays. In one of his experiments in early 1896, Becquerel exposed a crystal with uranium content to sunlight for a few hours. He then placed the uranium compound on a photographic plate and saw that the mineral produced its image on the plate. Becquerel believed that the absorbed energy of the sun was being released by the mineral in the form of x-rays.

Becquerel's efforts to repeat the experiment were delayed because of cloudy skies over Paris. He placed the mineral with the photographic plate in a drawer for a couple of days, waiting for sunny skies. When he opened the drawer to retrieve the mineral, expecting to see only weak images on the plate, he noticed that the mineral had left a new, clear, strong image on the plate. This was strange and unexpected! There was no source of energy to produce the image. Becquerel reported this to the Academy right away, saying that this was a new phenomenon and promising to continue this work and report back. He completed his experiments two months later and reported that the radiation came from the uranium itself, without a need for excitation by an external energy source.

In later experiments Becquerel found that the radiation emitted from uranium could be deflected by magnetic fields and consisted, therefore, of particles, unlike x-rays. The discovery of radioactivity by Becquerel in 1896 is one of the most unbelievable, exciting, and significant developments in all science and marked the beginning of modern physics, nuclear science, and the dawn of the nuclear age.

The term *radioactivity* was not invented by Becquerel but by Marie Curie, a remarkable scientist who was inspired by Becquerel's findings and was determined to study the new phenomena. Marie Sklodowska-Curie was born in Warsaw, Poland, in 1867, the youngest of five children in a family that struggled to make ends meet after the father lost his job. She was raised in Warsaw during the Russian occupation. Marie had a turbulent childhood, losing her sister, who died of typhoid fever, and her mother, who died of tuberculosis.

Marie eventually rejected her religious faith. She was a brilliant student and was awarded a gold medal as top student on graduation from high school. Her dreams of a higher education could not be fulfilled in Russian-dominated Poland, where girls were not allowed to attend university. She joined the Floating University, an underground educational organization that held secret educational sessions, providing Polish youth with an opportunity for an education that was free of the ideology of the governing foreign powers. The locale of the school sessions changed frequently to avoid detection by the Russians. Marie was taught physics, natural history, Polish history, and culture.

Working as a governess, the twenty-four-year-old Marie was able to save enough money for a train ticket to Paris and enrollment at the Sorbonne. She was able to move in with her sister, who was married and living in Paris. Marie enrolled as a student of physics in the Sorbonne in 1891. While at Sorbonne she was introduced to Pierre Curie, an instructor of physics and chemistry. Pierre was thrilled with Marie's drive and intellect and proposed marriage to her. Pierre's letter of proposal says the following: "It would be a beautiful thing to pass through life together hypnotized in our dreams: your dream for your country, our dream for humanity, our dream for science." Pierre Curie and Marie Sklodowska were married in 1895.

This was the beginning of an extraordinary scientific partnership. It was not long before the Curies turned their attention to Becquerel's discovery of the mysterious uranium radiation. They observed that the intensity of the radiation was in direct proportion to the uranium content in the sample, and nothing they did to the sample could stop the radiation. They concluded that the source of the radiation was the atom itself and began the search for other elements that showed similar activity. They studied pitchblende, an amorphous, black uranium oxide, a mineral that was well known for its uranium content. They found that this mineral was far more radioactive than uranium itself and concluded that it must contain other radioactive elements in addition to uranium. Using basic chemical refining methods, they isolated an element that had four hundred times the radioactivity of uranium. They named this new element polonium, in honor of Marie's native Poland. Further refining revealed the existence of another unknown element, which was later named radium, in recognition of its power of emitting energy in the form of rays. The Curies recognized the medical possibilities of this type of radiation, and in a presentation, Pierre Curie described the potential of radium in treating cancer.

In 1903 Marie Curie became the first woman in France to earn a doctoral degree in physics. Her doctoral thesis was about radiation, and her professors declared that it was the greatest contribution to science ever written. There were rumors of a Nobel prize for Marie Curie, but members of the French Academy of Sciences attributed the brilliance of Marie's work to her husband Pierre. Again, in those prejudicial times, it was unthinkable for a woman to be involved in professional scientific research, let alone be nominated for a Nobel prize. They lobbied quietly for the prize to be split between Becquerel and Pierre Curie. But Pierre had different ideas. He insisted that Marie had originated the research, planned all the experiments, and developed the theories explaining the phenomenon of radioactivity.

The 1903 Nobel prize in physics was finally awarded jointly to Henri Becquerel, Pierre Curie, and Marie Curie.

In order to appreciate the prejudicial attitude toward women scientists, we may reflect on the following quote from the Bible, restated by the president of the Swedish Academy during the awards ceremony: "It is not good that man should be alone; I will make a helpmeet for him." This is an unbelievable comment made by the highest official of the Swedish Academy, a comment that could not be made today.

In a tragic turn of events, Pierre Curie died in an accident in 1906 at the age of forty-seven, leaving Marie with two daughters. The joyful time of this incredible husband and wife team ended, and Marie Curie was faced with the gigantic task of moving forward with the research on radioactivity.

Marie was appointed to her husband's academic post, becoming the first woman professor at the Sorbonne in the 650-year history of the university. She edited and published all of Pierre Curie's unpublished work and published a massive work of her own, *Traite de Radioactivite*, a fundamental treatise on radioactivity and one of the most important books of science written in the twentieth century.

International recognition for Marie Curie was growing, and in 1911 she won the Nobel prize in chemistry, becoming the first person to win the Nobel prize twice. The Swedish Academy of Sciences had to overcome some opposition in awarding the prize. The opposition was a result of Marie's affair with physicist Paul Langevin, a married man estranged from his wife and a former student of Pierre Curie. The affair had resulted in a press scandal. In the xenophobic and prejudiced France of the early twentieth century, Marie Curie's academic opponents exploited the affair to block her membership in the Academy. Marie Curie died in 1934 at the age of sixty-six. Her death was attributed to a bone marrow disease believed to have been contracted as a result of her long-term exposure to radiation. In 1995 Marie and Pierre Curie's remains were enshrined in the Pantheon in Paris.

Marie Curie was a fascinating character and has been the subject of books, plays, and movies. Her legacy extends far beyond her pioneering scientific achievements. At a time when professional science was a man's world, her quest for scientific truth and her drive, persistence, and courage in overcoming so many social obstacles made her an ideal role model for young women. She donated most of her Nobel prize money to friends, family, students, and to the war effort during World War I. She insisted that monetary gifts and awards be given to the scientific institutions she was working with rather than to her. During the war she established a frontline x-ray service in the battlefields of France and Belgium, training the staff and driving the x-ray vans herself. Albert Einstein said that Marie Curie was probably the only person who could not be corrupted by money. The work of Henri Becquerel and Pierre and Marie Curie started a new scientific revolution and inspired scientists around the world to focus on the atom, the smallest indivisible piece of matter, as physicists believed at that time.

The existence of atoms had been recognized in antiquity around 400 BC, when the Greek philosopher Democritus first proposed the idea that everything is composed of atoms, which are physically indivisible. The word "atom" means "indivisible" and was coined by Greek philosophers. Conclusive evidence of the existence of molecules and atoms was provided by Scottish scientist Robert Brown, who in 1827 observed the random motion of particles suspended in a fluid (a liquid or a gas) resulting from their collision with the quick atoms or molecules in the liquid or gas. This phenomenon is now known as Brownian motion.

Eighteen centuries before Brown's observation, the Roman poet Lucretius wrote a remarkable scientific poem describing Brownian motion as follows:[13] "Observe what happens when sunbeams are admitted into a building and shed light on its shadowy places. You

---

13 Lucretius, *On the Nature of Things*, Project Gutenberg website, URL = http://www.gutenberg.org/files/785/785-h/785-h.htm

will see a multitude of tiny particles mingling in a multitude of ways... their dancing is an actual indication of underlying movements of matter that are hidden from our sight...It originates with the atoms which move of themselves [i.e., spontaneously]. Then those small compound bodies that are least removed from the impetus of the atoms are set in motion by the impact of their invisible blows and in turn cannon against slightly larger bodies. So the movement mounts up from the atoms and gradually emerges to the level of our senses, so that those bodies are in motion that we see in sunbeams, moved by blows that remain invisible."

The last decade of the nineteenth century brought an avalanche of scientific discoveries that placed the atom into central focus. Following Roentgen's discovery of x-rays in 1895, Becquerel discovered radioactivity in 1896, and Marie Curie discovered more radioactive elements in 1898. Between those two remarkable years, in 1897, a forty-one-year-old English physicist, Joseph John Thomson, discovered the electron and presented to a startled scientific community the idea that the atom is not the smallest piece of matter after all. Science was now inside the atom, and this was truly a new beginning in atomic and nuclear science.

Joseph John Thomson, or simply "J. J.," as he was affectionately called, was born in 1856 in Cheetham Hill, a suburb of Manchester, England. After completing his early education, he was admitted to the University of Manchester at the age of fourteen, an unusually early age for university admission. At the age of twenty-eight, he became Cavendish Professor of physics at the University of Cambridge. The post gave Thomson supervision of the famed Cavendish laboratory, a physics lab started in 1871 with James Clerk Maxwell as the first Cavendish professor. At Cambridge, Thomson turned out to be a superb educator. He took an active interest in the work of young researchers, checking their progress on a daily basis and often making suggestions for improvements. Seven of his research assistants and Thomson's own son won Nobel prizes in physics. Thomson is probably the greatest educator of modern time if we only count the number of Nobel laureates among his students.

Thomson turned his scientific interests to the study of cathode rays. We all know these from our familiar old style television sets, which used cathode ray tubes. The cathode in the tube produced an electron beam, which was deflected by electric or magnetic fields to create images on the screen. When Thomson applied an electric field to the cathode, he noticed a glow in the tube. He was able to deflect the glow with electric and magnetic fields and determined that it consisted of a stream of particles. With further experiments he was able to determine the ratio of electric charge to mass, and his calculations showed that these particles were at least one thousand times smaller than hydrogen, the smallest known atom. He called them corpuscles, just as Newton had done two hundred years before him. They were later named electrons. The atom was no longer the smallest unit of matter. Smaller particles were residing inside the atom. Thomson presented three hypotheses about the nature of these particles. First, cathode rays consist of charged particles. Second, the particles are constituents of the atom. Third, the particles are the *only* constituents of the atom.

The second and third hypotheses were especially controversial among scientists, who generally met Thomson's speculation with skepticism. In fact, we know now that his third hypothesis is false. Scientific acceptance was gradual, and at the same time, other scientists embarked on new research focusing on atomic structure. In 1904 Thomson proposed a model of the atom, most often called the "plum pudding model," sometimes known as the "blueberry muffin model" or the "raisin cake model."

In the plum pudding model, the atom is composed of negative charged particles surrounded by a cloud of positive charge that balances the negative charges of the particles. The particles are the raisins, and the positively charged cloud is the pudding. In this model the particles were free to rotate within the cloud of positive charge. For his discovery of the electron and his work on the conduction of electricity in gases, Thomson was awarded the 1906 Nobel prize in physics and was knighted in 1908.

Research in physics was thriving, and Thomson's plum pudding model would not last very long. New Zealander scientist Ernest Rutherford disproved the model in 1911. Born in New Zealand in 1871, Rutherford completed his early education in the public school system of his native country. At the age of twenty-four, he won a scholarship to study at Cambridge. This gave him an opportunity to do research under Thomson at Cavendish. Rutherford's talents were quickly recognized by Thomson. The two scientists worked jointly on research studies, particularly in the study of x-rays, which had just been discovered by Roentgen and were a hot topic among research scientists. Their research focused on the way that x-rays changed the conductivity of gases. They wrote a classic paper on ionization, which is the breaking of atoms and molecules into positive and negative parts called ions. Rutherford's attention soon turned to radiation produced by radioactive substances such as uranium and its compounds, the type of radiation discovered by Becquerel and the Curies. Rutherford placed uranium samples near foil and observed that some of the Becquerel radiation was blocked by the foil while the rest penetrated quite easily. He theorized that these were two different types of radiation and called them "alpha" and "beta," from the first two letters of the Greek alphabet. It was determined later that an alpha particle is the same as a helium atom without its electrons; it is the nucleus of an ordinary helium atom. The beta particle is the same as an electron or its positive version, the positron, a particle that was discovered in cosmic radiation much later.

In 1898 Rutherford was offered a professorship at McGill University in Montreal, Canada. The university had one of the best physics laboratories in the world, and Rutherford accepted the offer. While at McGill he continued his work on radioactive bodies; and together with English chemist Frederick Soddy, he developed the theory on the disintegration of radioactive elements. It is Rutherford who coined the term "half-life," which is the time required for the reduction of a radioactive element's mass to half of its present quantity. Rutherford went back to Cambridge in 1907 and was awarded the 1908 Nobel prize in chemistry for his work on the disintegration of radioactive substances. Joking about the fact that the prize was in chemistry rather than physics, Rutherford said in his acceptance speech that he had seen many transformations in his studies, but never one so rapid as his own from physicist to chemist!

Rutherford's research efforts were far from over. In one of his experiments in 1911, he directed a beam of alpha particles to a thin gold foil and studied their deflections. In a breakthrough of unprecedented proportions, he proved that almost all the mass of an atom is concentrated in the nucleus. Rutherford is justly regarded as the father of nuclear science. In two exciting decades, Becquerel, the Curies, Thomson, and Rutherford became the undisputed protagonists of a new scientific revolution, one we might call "the atom is not the smallest unit of matter." Scientists now had the huge task of trying to look inside the atom. The nuclear age had begun.

One of the most intriguing scientific stories of the nineteenth century is the development of our knowledge about light. Newton had proposed in 1675 that light consists of single, infinitesimally small particles, which he called corpuscles, from the Latin corpus (body). These particles were thought to be emitted in all directions from a specific source. The particle theory was able to explain some optical phenomena by assuming that light was attracted to mass by gravitational forces.

Newton's theory gained wider acceptance among scientists over the competing wave theory championed most prominently by Dutch scientist Christiaan Huygens. Born in The Hague in 1629 into an important, well-connected family, Huygens studied law and mathematics at the University of Leiden and the College of Breda. He showed great talent as a mathematician, and his publications gained him wide recognition. He was also interested in astronomy and discovered one of Saturn's satellites. His most important contribution is the wave theory of light.

According to the wave theory, light is emitted in all directions as a series of waves in a medium called ether. As light can travel in a vacuum, it was assumed that vacuum and empty space are filled with ether. This was the big problem with the wave theory. It required the existence of this transmitting medium, an existence that could not be verified by any experiment. We know that sound waves cannot travel in a vacuum. We also know that ocean waves are movements of water molecules. You cannot have a wave of nothing. Waving is a behavior, and you need a subject that exhibits that behavior.

This weakness of the wave theory was the main reason Huygens had to invent ether and why Newton's particle theory dominated the scientific community during most of the eighteenth century. Then came an experiment in the late nineteenth century that provided strong evidence against the existence of ether. It was performed by two American physicists, Albert Michelson and Edward Morley in 1887 at the Case Western Reserve University in Cleveland. But this experiment required a fairly accurate knowledge of the speed of light.

Measuring the speed of light is no fun unless you are a Danish astronomer named Ole Roemer working at the Paris observatory. Roemer observed in 1676 that as the earth and the planet Jupiter moved in their orbits around the sun, the distance between them varied. One of Jupiter's moons, Io, was eclipsed by Jupiter at regular intervals. Roemer's observations over several months showed that the eclipses lagged more and more each time, and then they suddenly appeared to pick up again. Roemer attributed this to the varying distance between Jupiter and the earth, which was due to their orbital movements around the sun. Roemer's calculations finally showed that light from the sun took about eleven minutes to reach the earth. In spite of Roemer's good efforts, this was a significant miscalculation. It was corrected by Newton's observations a few years later, which showed the time to be closer to seven or eight minutes. Newton's estimate was essentially the correct value. Many years later, Michelson made a much more accurate measurement that showed the speed of light to be 186,355 miles per second. That is extremely close to our current knowledge of 186,282 miles per second. We generally refer to the speed of light as 186,000 miles per second or 300,000 km per second.

Armed with an accurate knowledge of the speed of light, Michelson and Morley proceeded with their experiment to detect and measure the movement of the earth relative to ether. It was typically believed that ether was rather static, but the earth moves around the sun at a fairly high speed, nineteen miles per second or roughly 67,000 miles per hour. There would be a relative speed of this order of magnitude between ether and the earth. Michelson and Morley thought they should be able to measure this relative speed.

Think about when we swim in a river. It takes longer to swim upstream than it does to swim downstream. The speed of the river's flow adds to our swimming speed, and we cover a given distance in much less time than swimming against the river flow. Michelson and Morley thought that if they could send a beam of light to a half-silvered mirror at an angle, the beam would split into two beams, one reflected from the mirror and the other going through. With a suitable arrangement of mirrors, they would force the two beams to travel in opposite directions and then hit the same target and reunite at the target. As one beam traveled in the direction of the earth's motion against ether, and the other beam traveled in the opposite direction, the two beams would not arrive at the target at exactly the same instant. This phase shift would be visible as an interference pattern on the target screen.

The experiment was bold and required a great deal of accuracy, but Michelson's experimental genius was up to the task. Michelson planned the experiment and the apparatus for a long time, spanning at least a couple of years. The planning was so intense that Michelson suffered a nervous breakdown during this time. With Morley's help he finally set up an apparatus of astonishing accuracy, consisting of a light source, a half-silvered mirror, two more mirrors, and a screen as a target for the final beam of light. The entire apparatus was mounted on a turntable so that all possible directions of ether wind could be tested.

The experiment produced unexpected results. No interference was observed on the target screen. The experiment was repeated with improved accuracy many times, but always with the same results. The conclusion drawn was that the speed of light is always the same regardless of the direction that light was traveling in. If we go back to our analogy, our swimmer took exactly the same time to cover the same distance swimming upstream as swimming downstream.

This was a major setback for the wave theory of light, and Michelson and Morley's experiment became known as the most famous failed experiment in the history of science. It wasn't really a failure, as its findings opened the path to Einstein's relativity theory as we shall see later. But the wave theory was back to square one. Something that does all that waving still needed to be found for the theory to have any scientific status.

The validity of the Michelson and Morley's experiment rests on the assumption that ether is static, fixed in space, and can be regarded as an absolute frame of reference. On the other hand, we know that everything in the universe travels at phenomenal speeds. The sun, along with the entire solar system, travels at roughly forty-three thousand miles per hour on its orbit around the center of the galaxy. The proposition that ether (or anything else, for that matter) is static against the moving earth is not so obvious—and certainly not intuitive.

The analogy with the river swimmer is not a great analogy, but we have added it here as it has been used widely in the literature. The swimmer is fighting the friction of water and the force of gravity when swimming upstream. Light photons may not be experiencing any such forces from ether. Also, physicists agree that the speed of light is constant and is not affected by the relative velocity between the source of light and the observer, and yet the success of Michelson and Morley's experiment depended on a speed difference based on which direction light traveled against ether. There are contradictions and fallacies that lead to the conclusion that the Michelson-Morley experiment did not confirm or disprove anything, as it failed to fulfill the logical requirements of scientific methodology.

The flaws of the experiment did not prevent the scientific community from overinterpreting the result. This is a classic case of groupthink, as we shall see later. Scientists were eager to abandon the uncomfortable ether, and many embraced the results of the experiment without sufficient scrutiny. Others were not as convinced. Twenty-two years after the Michelson-Morley experiment, English physicist J. J. Thomson, discoverer of the electron, proclaimed, "The ether is not an imaginary creation of the speculative philosopher. It is as essential to us as the air we breathe."

It wouldn't be long before a new theory would come forward, a theory that would not need the existence of a transmitting medium for the propagation of light. It was a theory that would satisfy everyone and would begin the modern era in man's understanding of the universe.

Pragmatism in America

Peirce, James, and Dewey

A NEW PHILOSOPHICAL CURRENT began to develop as nineteenth-century America was coming of age, culturally and intellectually. The undisputed founder of the new philosophy was Charles Sanders Peirce, one of the most original minds of the century and, in the words of Bertrand Russell, "the greatest American thinker ever."[14] Umberto Eco's assessment is no less approving: "Peirce was the greatest American philosopher of the turn of the century and beyond doubt one of the greatest thinkers of his time."[15]

Peirce was born in 1839 in Cambridge, Massachusetts, where his father was a professor of mathematics and astronomy at Harvard. His lifelong fascination with philosophy and especially logic started when he was twelve and read his brother's textbook, a treatise titled *Elements of Logic* by Richard Whately, a prominent English philosopher

14 Bertrand Russell, *Wisdom of the West*, (New York: Doubleday, 1959), 276.

15 Umberto Eco, introduction in C. K. Ogden and I. A. Richards's *The Meaning of Meaning*, 4th ed. (San Diego: Harcourt, 1989).

and theologian. Peirce studied chemistry at Harvard but was never employed there due to an unfavorable opinion held by one of his professors. After working in various scientific jobs, he was appointed lecturer at Johns Hopkins University in Baltimore. A true polymathic researcher, Peirce published articles on mathematics, logic, physics, geodesy, spectroscopy, astronomy, psychology, anthropology, history, and economics. He was an intellect of incredibly wide range, a supreme polymath in the Aristotelian sense.

Amazingly, Peirce did not publish any philosophical works. We know of his philosophical theories from technical articles he published in various journals. Much of what Peirce wrote remains in manuscript form, unpublished and in complete disarray. The interested reader might want to consult the *Cambridge Companion to Peirce* or choose from a plethora of secondary sources. Peirce lived the last twenty-five years of his life in dire poverty, as he had lost his job and was considered unemployable in academia due to his alleged erratic behavior. He died in 1914, and his death went largely unnoticed in the philosophical community.

Peirce's method of scientific inquiry has three stages: abduction, deduction, and inference. Abduction refers to the generation of a hypothesis that may explain a particular phenomenon. Deduction is the stage where propositions are formed on how the hypothesis will be tested. Inference is the stage of experimentation conducted in order to test the hypothesis.

This sounds quite simple—and actually quite reasonable, too. We know that this is the process frequently encountered, in some form or another, not only in scientific work but also in our everyday reasoning. Peirce is the first person who coined the term "abduction," but Peirce's use of the term is different from the way we use it today. In modern usage, abduction is part of the inference stage, where we make assessments of hypotheses and theories. In Peirce, abduction is part of the first stage of inquiry, where we develop hypotheses and theories to be tested at a later stage.

One of the older issues in philosophy concerns the deterministic versus indeterministic nature of the world and also of human behavior. We saw in our discussion of Aristotle how chance played a role in his explanation of physical phenomena. Certain phenomena were due to specific causes, but other phenomena were not. An undetermined cause called *tyche*, which is the Greek word for chance, was the driving force.

Aristotle's idea of randomness is one of the ruling forces in quantum physics today but was left unexploited for centuries until Peirce came along. Before we can understand the role of chance in the natural world, we will need to clarify the two philosophical concepts of determinism and indeterminism. Determinism is the position that for every event, conditions exist that could cause no other event. It does not necessarily mean that everything that happens is predetermined. Determinism is not the same thing as fatalism. Determinism simply means that events are caused by a specific combination of conditions that could have no other outcome.

Let us consider an imaginary series of events before we get into Peirce's position on the subject: John goes to a lottery outlet to buy a lottery ticket. He just makes the deadline and buys the last ticket sold. By doing so, John has a very small chance of winning the jackpot. On his way to the same lottery outlet, Peter misses the bus because he slips on the wet ground, and so he arrives with the next bus a few seconds after the outlet closes. Peter is unable to buy a ticket and has zero chance of winning the jackpot. The draw is held, and John's ticket wins the jackpot.

We have here a series of events, some deterministic and some not. The two facts that (1) John makes the deadline and (2) his possession of the ticket are related deterministically. If you make the deadline, you are assured of a ticket. Also, the fact that John possesses a ticket is deterministically related to the fact that he has a chance of winning. But the facts of John's possession of the ticket and his winning the jackpot are related by chance and are not deterministic. As for Peter, his story is full of chance events except one: his failure to buy a ticket and his zero chance to win are related deterministically.

We see, therefore, that both determinism and chance are everywhere when events of human behavior are concerned. We will opt for a liberal interpretation of determinism that allows for chance to have a role. For example, some of the conditions that lead mathematically to a specific event may themselves be outcomes of chance.

What about the natural world? Why do we use statistical and probabilistic methods to understand certain phenomena? Is it because of our lack of a complete analytical understanding of the phenomena, or is it because the phenomena themselves are subject to chance? Is there an irreducible element of chance that is present in most or all processes?

Peirce's philosophy is on the camp opposite to determinism. The Enlightenment in Europe had brought powerful forces of determinism into eighteenth-century philosophy. Peirce urged against these currents, saying that there was no scientific evidence that pointed to determinism, and, in fact, there was plenty of evidence against it.

Peirce wrote an article in 1878, in which he summed up the basic principle of pragmatism: "Consider what effects, that might conceivably have practical bearings, we conceive the object of our conception to have. Then, our conception of these effects is the whole of our conception of the object."[16] In other words, the practical consequences of an object define the concept. The very meaning of concepts is most adequately clarified in terms of their conceivable practical consequences.

The concept of practicality in Peirce is somewhat broader than what we mean by the term in everyday language. In addition to practicality of usage, it includes any consequences of the object that are apt to affect conduct. Peirce's view is similar to that of Einstein, who

16 Charles Sanders Peirce, "How to Make Our Ideas Clear," *Popular Science Monthly* Vol. 12 (1878): 293.

held that the whole meaning of a physical concept is determined by an exact method of measuring it.

Let us think of an example. When an infant sees a table for the very first time, the concept "table" is formed in his or her mind. The concept has all the elements of that first table—its color, size, type of material, number of legs. When the infant sees another table, the process of abstraction begins. The concept "table" is reduced to the common features of the two tables. The final concept, after the experience of several tables with and without people using them, has none of the particular features, such as color, size, and so on. But it does have the feature of usage. When we think of the concept "table," we do not think of a specific color or style but rather of what it can do for us. That practicality entirely defines the concept "table." The idea "table" is now a conceptualized form, which is really what Plato had in mind when he developed his idea of Forms. The difference is, of course, that Plato's forms preexisted in the mind and were the genesis rather than the outcomes of reality.

In the same article, Peirce defines his concept of "truth": "Different minds may set out with the most antagonistic views, but the progress of investigation carries them by a force outside of themselves to one and the same conclusion. This activity of thought by which we are carried, not where we wish, but to a foreordained goal, is like the operation of destiny. No modification of the point of view taken, no selection of other facts for study, no natural bent of mind even, can enable a man to escape the predestinate opinion. This great law is embodied in the conception of truth and reality. The opinion which is fated to be ultimately agreed to by all who investigate, is what we mean by the truth, and the object represented in this opinion is the real. That is the way I would explain reality."[17] That is, reality and truth are determined by the consensus of those who investigate. The force of consensus is beyond the power of each individual. Peirce's view is

17 Charles Sanders Peirce, "How to Make Our Ideas Clear," *Popular Science Monthly* Vol. 12 (1878): 286-302.

based on the assumption that there is only one reality. The common elements of individual human perceptions form a consensus, which is the best representation of truth.

In our review of scientific discoveries in this book, we have seen that a theory does not reach the status of scientific truth until it is widely accepted by the scientific community. We will certainly see more evidence of this later on. It is a democratic process, after all. The scientific community "votes" and accepts that the theory is a true representation of the phenomenon. Peirce's consensus view is quite consistent with what we have seen in our study of the evolution of scientific knowledge.

Peirce was a true Renaissance man, and we have only touched one aspect of his philosophy of the mind. Charles Peirce and William James are often said to be the founders of the American renaissance. English mathematician and philosopher A. N. Whitehead said that James is the analogue to Plato and Peirce to Aristotle.

William James was born in New York City in 1842 to a wealthy, intellectual, cosmopolitan family that was constantly a major cultural focus and a subject of continuing interest to historians and critics. William was the brother of notable writer Henry James, who was one year his junior. William received an eclectic education and learned French and German. The family's wealth allowed them to travel to Europe frequently, with the aim of gaining a world perspective and the best possible education for the children. After an early flirtation with art, William's interests were focused on scientific studies, and he enrolled at Harvard to study medicine, physiology, and biology. He graduated in 1869 but never practiced the medical profession, as his interests had turned to psychology and philosophy. His financial disposition allowed him many alternative choices of what to do with his life. This kind of freedom and the resulting indecisiveness caused him frequent spells of depression, on top of his delicate health, chronic back pain, bad eyesight, and insomnia. By the time James reached middle age, he still didn't know what to do with his life, and he became suicidal. He was troubled by the philosophies of determinism; if everything is determined by prior events, then what is the point of planning anything in one's life?

In the 1870s James was appointed to various academic posts at Harvard, reaching the status of full professor in 1885. Five years later James wrote his monumental two-volume *Principles of Psychology*, which included his famous account of free will. His last major work was titled simply *Pragmatism* and included a series of lectures on his theory of truth, meaning, and knowledge. This work established James as a leader in the philosophical movement of pragmatism and is regarded as the most influential book of American philosophy. James died of heart failure at the age of sixty-eight.

James's works are much more readable and lighter on technical terminology than those of his close friend Charles Sanders Peirce. James borrowed the term "Pragmatism" from Peirce, who believed that our understanding of truth and meaning is guided by pragmatic concerns. James clarified and popularized Peirce's theory. He begins his famous sixth lecture ⊠n pragmatism with a dictionary analysis of truth as "agreement with reality," a definition that, on the surface, is widely agreed upon. The contentious words are "agreement" and "reality." Intellectualists think that ideas are copies of what is fixed and independent of us. The pragmatist, on the other hand, takes a more dynamic and practical view. A true idea is one we can incorporate into our thinking in a way that the idea can be experientially verified and validated by experience. In his *Pragmatism: A New Name for Some Old Ways of Thinking*, James writes:

> Pragmatism asks its usual question. 'Grant an idea or belief to be true, what concrete difference will its being true make in anyone's actual life? How will the truth be realized? What experiences will be different from those which would obtain if the belief were false? What, in short, is the truth's cash-value in experiential terms?' The moment pragmatism

asks this question, it sees the answer: TRUE IDEAS ARE THOSE THAT WE CAN ASSIMILATE, VALIDATE, CORROBORATE AND VERIFY. FALSE IDEAS ARE THOSE THAT WE CANNOT. That is the practical difference it makes to us to have true ideas; that, therefore, is the meaning of truth, for it is all that truth is known as. This thesis is what I have to defend. The truth of an idea is not a stagnant property inherent in it. Truth HAPPENS to an idea. It BECOMES true, is MADE true by events. Its verity is in fact an event, a process: the process namely of its verifying itself, its veri-FICATION. Its validity is the process of its valid-ATION.[18]

Farther down in the same text, James explains the notion of usefulness as a necessary condition of Truth:

The importance to human life of having true beliefs about matters of fact is a thing too notorious. We live in a world of realities that can be infinitely useful or infinitely harmful. Ideas that tell us which of them to expect count as the true ideas in all this primary sphere of verification, and the pursuit of such ideas is a primary human duty. The possession of truth, so far from being here an end in itself, is only a preliminary means towards other vital satisfactions. If I am lost in the woods and starved, and find what looks like a cow path, it is of the utmost importance that I should think of a human habitation at the end of it, for if I do so and follow it, I save myself. The true thought is useful here because the house which is its object is useful. The practical value of true ideas is thus primarily derived from the practical importance of their objects to us.

---

18 The capitalizations are in William James's original text.

In 1904 James published an essay titled "Does Consciousness Exist?", wherein he explained his theory of radical empiricism. In old philosophy there is a distinction between subject and object. There is an occurrence of knowing and the subject is the knower, while the object is the thing to be known. The subject is the mind or soul, and the object can be a material object, an idea, another mind, or even the subject itself. There is a dualism of mind and matter, a world consisting of two distinct, ontologically separate categories. In science, the subject is the observer, and the object is the phenomenon to be observed.

James does not accept that this dualism of subject and object is fundamental. Experience is not just a stream of data sensed by our senses and imprinted in our minds as perceptions, concepts, and ideas. It is a complex process of interaction shaped by what the objects mean to us and by the causal relationships with other phenomena. The interaction is dynamic and open to new experiences, new meanings and new causal relationships.

A truly original American school of philosophy, Pragmatism found another major proponent in John Dewey. Widely regarded as the leading American philosopher of the twentieth century, Dewey was born in 1859, only seventeen years after the birth of William James, but he lived well into the mid-twentieth century. Dewey was born in Burlington, Vermont, and studied at the University of Vermont and Johns Hopkins University. He held academic posts in Michigan, Chicago, and finally at Columbia University in New York, where he spent most of his life. During his student days, Dewey became aware of evolutionary theory, which would influence his thinking for the rest of his life. In the world before Darwin, organisms and species were seen as well-defined entities, as created by the Creator. By the time Dewey was a young university student, Darwinian evolution had been accepted by the scientific community and the general public. The idea that all species had descended over time from common ancestors through natural selection, adaptation, and iterative change had an immense impact on young Dewey and became central to his view of the natural world and his conception of human nature.

Dewey didn't write any major works until he was fifty, when he wrote his book *How We Think*. In this work Dewey reveals his ideas on inference, understanding, and empirical and scientific thinking, all with a dual purpose: to understand the nature of knowledge and to apply this understanding in shaping his own teaching methods. Dewey's interest in education was so strong that he wrote on the subject as much as on philosophy, psychology, and social theory. In the first half of the twentieth century, Dewey became the leading education reformer in America.

The theory of knowledge is a central focus in Dewey's philosophical work. He rejects the term "epistemology," preferring the "theory of inquiry" and "experimental logic" as more descriptive of his own work. Dewey believes that knowledge develops out of a continuing interaction between the mind and the environment and the mind has practical instrumentality in the guidance of that interaction, aiming at restructuring the environment and its practical consequences. Darwin's influence is quite evident in Dewey's naturalistic approach, and so are the influences of Peirce and James. Dewey rejects the old mind-matter dualism and proposes an alternative view. In its interaction with the world, the mind energizes, coordinates, and integrates sensory and motor responses. Learning is not a process of passively imprinting impressions upon the mind but rather an active manipulation of the environment.

The continuing interaction of mind and object takes the form of an amphidromous sequence of deductive and inductive reasoning. In Dewey's own words:

> There is thus a double movement in all reflection: a movement from the given partial and confused data to a suggested comprehensive (or inclusive) entire situation; and back from this suggested whole, which as suggested is a meaning, an

idea, to the particular facts, so as to connect these with one another and with additional facts to which the suggestion has directed attention. Roughly speaking, the first of these movements is inductive; the second deductive. A complete act of thought involves both: it involves, that is, a fruitful interaction of observed (or recollected) particular considerations and of inclusive and far-reaching (general) meanings.[19]

In other words, Dewey proposes using deductive reasoning to develop a theory, followed by inductive reasoning to support it. Dewey believes that science is the most perfect type of knowledge because it uses causal definitions. Dewey's scientific method consists of five logical steps: (1) identify and define the problem, (2) make a hypothesis of why the problem exists, (3) collect and analyze data, (4) formulate conclusions, and (5) apply conclusions to the original hypothesis.

Dewey's theory of knowledge has elements of fallibilism. That is a fancy word for something rather simple, but we know by now that philosophers like to brighten up their dark subjects with fancy words. Fallibilism is just the attitude of being open to new evidence that contradicts a previously held concept or theory and requires its revision. Scientists take this for granted in their daily work and actively look for contradictions and revisions. Dewey's mind-matter interaction obviously has an inherent ability of constant review and revision. But this is a good time to get out of the dark labyrinths of philosophy and out into the bright, real world of science.

---

19 John Dewey, *How we think*, (Lexington, Mass: D.C. Heath, 1910): 79.

The Stochastic Universe

A Discrete Revolution

T HE MIRACULOUS LAST decade of the nineteenth century brought scientific focus into the interior of the atom, which was no longer the smallest unit of matter. The twentieth century was destined to open with a bang when, in 1900, German physicist Max Planck made a pioneering proposal that would begin the quantum era in our modern understanding of the microcosm. Planck was born in 1858 in Kiel, Germany, the sixth child of a traditional, intellectual family. His father was a law professor at the University of Kiel. The young Max grew up in a family environment dedicated to church and state, academic excellence, conservatism, idealism, and generosity.

Max excelled in all subjects at school, but at a young age, he developed a strong interest in physics, astronomy, and mathematics. His polymathy extended into philosophy, theology, and music. He played the piano, organ, and cello and composed songs and operas. When Max was nine years old, the family moved to Munich, where the elder Planck had been appointed to a professorial chair. At the age of sixteen, Max enrolled in the university, where the professor of physics advised him not to go into physics because everything in physics had already been discovered. Planck replied that he did not necessarily want to discover anything new but only to understand the fundamentals. What a wonderful response from someone who was going to start one of the greatest scientific revolutions of all time!

In 1889 Planck moved to Berlin, where he became a full professor in 1892. He was now married with four children, and the Planck home in Berlin became a social and cultural center for academics and scientists, including Albert Einstein. Planck's early work was in thermodynamics, an interest acquired when he was a student of eminent physicist Gustav Kirchhoff in Berlin. He published papers on entropy and thermoelectricity. He was also interested in radiations and believed their origin was electromagnetic. His work focused on the relationship between the energy and the frequency of radiation, which he determined to be $E = hv$, where E is the energy of one photon of radiation, v is its frequency, and h is a constant number, now known universally as Planck's constant.

Planck's insight, powerful logic, and mathematical ability made him the supreme theoretical physicist. The road to scientific truth is not a single path. It is a labyrinth of parallel, circular, and diagonal paths that you need to walk to get to a certain scientific truth. Sometimes they will not get you anywhere except back to the starting point. One such path is to observe a phenomenon, develop a theory that explains it, and then perform an experiment to test the theory. Planck was not an experimentalist. He used the results from experiments performed by others, and with his formidable theoretical mind, he developed a theory that explained the results. This is the purest form of inductive reasoning.

In the last decade of the nineteenth century, Planck began to investigate the problem of blackbody radiation, first posed by Kirchhoff some forty years earlier. The term "blackbody" refers to a physical body that absorbs all radiation, such as light, and does not reflect any of it back. The blackbody will, however, emit radiation when heated to a high temperature. We know that any solid heated above a certain high temperature will begin to glow—that is, emit visible light. At lower temperatures a solid still emits radiation, but its intensity in the visible-light region is too weak to be seen. The hotter the body, the brighter the glow. From our discussion of the light spectrum earlier in the book, we know that a brighter glow indicates light of higher frequency. There is a relationship, therefore, between the energy received by the body in the form of heat and the frequency of the radiation emitted.

Planck was determined to understand and quantify this relationship. He used established methods that we would now call classical physics, but in Planck's time, they were the avant-garde of modern physics. Planck's theoretical results from the classical methods did not match the experimental observations. He could not match the theory with the experimental results unless he assumed that radiation of energy is not a continuous wave but a discrete emission, an emission of small, discrete bits of energy, similar to bullets fired from a gun in quick succession. These discrete bits of energy were what we know today as quanta or photons.

The scientific community, including Planck himself, did not appreciate the full importance of this new, revolutionary idea. Initially there was a great deal of reluctance toward accepting Planck's theory. Nobody really believed that photons existed. Acceptance of the theory's radical implications was gradual and took time as every experiment that was devised to test the theory produced results that were predicted by the theory. This was the beginning of quantum physics, the beginning of an entirely different understanding of the microcosm. Planck laid the foundation upon which Bohr, Einstein, Schroedinger, and Heisenberg would later build the entire structure of modern physics.

Planck's theory of energy quanta is a milestone in the history of physics and was recognized with the Nobel prize in physics in 1918.

But the second half of Planck's life was marked by a series of tragic events. His wife, Marie Merck, a childhood friend, died in 1909 after twenty-two years of marriage, leaving him with two sons and twin daughters. He married her cousin Marga von Hoesslin two years later. Planck's oldest son was killed in World War I. His two daughters both died in childbirth. Another son was executed by the Nazis as a conspirator in an assassination attempt against Hitler. Planck himself became active during the rise of the Nazi regime in efforts to stop the persecution of Jewish scientists, who had been fleeing the country and its universities. The war and Planck's personal tragedies destroyed much of his will to live. Planck died in Gottingen in 1947.

By the end of the first decade of the twentieth century, Planck's extraordinary quanta theory, in spite of its slow acceptance, had become a necessary foundation for most research in physics. The discrete nature of electromagnetic radiation could no longer be ignored. Planck's theory, combined with Rutherford's discovery of the atomic nucleus in 1911, pointed to a new path of scientific discovery, a new model of the atom, one that would be far more credible than the existing concepts. Such a model was proposed in 1913 by Niels Bohr, a Danish physicist.

Born in Copenhagen in 1885, Bohr came from an intellectual, upper-class family. His father was a professor of physiology at the University of Copenhagen, and his mother came from a prominent Jewish family with connections to banking and parliamentary circles. After the completion of his secondary education, Bohr enrolled at the University of Copenhagen, where he studied physics, mathematics, astronomy, and philosophy. Following his graduation, Bohr was invited by Rutherford to conduct postdoctoral research at Victoria University in Manchester, England. Bohr embarked on a study of the structure of atoms based on Rutherford's discovery of the atomic nucleus and Planck's radical theory of energy quanta. The outcome of this research was a new model of the atom, published by Bohr in 1913.

Bohr's model is a perfect synthesis of Rutherford's atomic model and Planck's quantum theory. It is a planetary model, with electrons orbiting the nucleus much like the way planets orbit around the sun. There is a dense central region called the nucleus. An electron can move to a higher orbit if it receives energy—in the form of heat, for example. It can also move to a lower orbit by releasing energy in the form of a discrete energy packet, a quantum.

The similarity of Bohr's atomic model to the planetary model of the solar system is striking. The reason for the similarity is that the electric attraction between two electric charges and the gravitational attraction between two masses have the same mathematical form. The electric attraction is proportional to the product of the two charges divided by the square of the distance between them. This is known as Coulomb's law, discovered in 1785 by French physicist Charles Augustin de Coulomb. The gravitational attraction between two masses, discovered by Newton, has exactly the same form. It is proportional to the product of the two masses divided by the square of the distance between them.

Is this pure coincidence, or is there some larger law of the universe that could explain both types of attraction forces in one unified theory? Some stars are very large, and some are smaller than our sun, just like some atomic nuclei are large, and some are much smaller. Are atoms tiny solar systems? Are solar systems the atoms of a much larger entity? It is quite possible that such questions will never be answered. The size of the human scale compared to the cosmological scale most likely means that there are limits to our capacity of scientific knowledge.

In the meantime, physics can be fun, as we are probably nowhere near those limits yet. The Bohr model has a basic feature that makes it quite different from Rutherford's model. The energy of the electrons can only take discrete values in accordance with Planck's law. As the electron releases a quantum of energy, it must move to a lower orbit. Since the energy is quantized, the orbits are quantized as well.

Only certain orbits with certain radii are allowed. Other orbits simply do not exist. Electrons cannot take small steps. They have to jump between allowable orbits. It is all or nothing. That is probably why our daily folklore has been enriched with the expression "quantum leap." You have to gather enough energy to make the quantum leap, or else you stay where you are.

The lowest energy state occupies the lowest orbit, the one closest to the nucleus. This is called the ground state and has a quantum number of one. The next higher orbit has a quantum number of two, and so on. These higher states are also known as excited states. So we have the first excited state, the second excited state, and so on. The highest excited state is an ionization state. The electron at that outer orbit is no longer bound to the atom and soon or later will escape, creating an electron deficiency, which transforms our atom into an ion. When the electron jumps from one orbit to the next lower orbit, it emits a photon with a fairly low frequency, and we see this as a reddish glow. When it jumps down two or three orbits from an outer orbit, it emits a photon of higher frequency, and we see this as a greenish or bluish glow.

Bohr's work on the structure of atoms and the energy emanating from them won him the Nobel prize in physics in 1922. Bohr became so popular in his country that the Carlsberg brewery gave him, as a gift, a house right next to the brewery, with a pipe connection between the house and the brewery. Bohr could draw as much free beer as he liked! We do not know whether Bohr showed greater appreciation for his direct connection to Carlsberg's beer than his Nobel prize. One might think that both rewards were quite welcome in the Bohr household.

Because of his Jewish background, Bohr was forced to flee to Sweden in 1943 during the German occupation of Denmark. In fact, he managed to escape just before he was about to be arrested. Bohr was familiar with German nuclear research through his friendships with German scientists such as Werner Heisenberg. From Sweden he traveled to London, where he became involved in the British nuclear weapons program. At about the same time, he became a senior consultant to the Manhattan Project, which was the American nuclear weapon research program. He worked in the top-secret Los Alamos laboratory in New Mexico, where, for security reasons, he was known under the assumed name of Nicholas Baker. Bohr was deeply concerned about a nuclear arms race and believed that atomic secrets should be shared by the international scientific community.

Bohr met with Franklin Roosevelt as well as Winston Churchill but failed to convince them that atomic research should be shared worldwide. Churchill was strongly opposed to sharing any research with Soviet Russia and considered Bohr an unstable person and a dangerous security risk. After the war Bohr returned to Denmark, where he was decorated once again for his scientific achievements. He was appointed to serve in various public roles, such as the presidency of the Royal Danish Academy, and became a strong advocate for the peaceful use of nuclear energy. Bohr died in Copenhagen in 1962 at the age of seventy-seven.

The research into the inner structure of the atom had to continue. Bohr's model was great when applied to a simple atom, such as the hydrogen atom, but failed to explain the behavior of more complex atoms. The Bohr model was able to explain the light spectra emitted by the single hydrogen electron as it jumped orbits and released photons of visible light. The model was not successful in predicting light spectra of atoms with more than one electron. But the model was a good starting point toward a more comprehensive and a much more complex atomic structure.

None of this diminishes the value of Bohr's achievement. His atom explains why materials can emit light and does so in full agreement with Planck's law of quanta. Planck's and Bohr's achievements made it clear that the laws of classical mechanics were not able to explain the microcosm. They were derived from human observations of earthly, macroscopic phenomena, while the microcosm moved at speeds and frequencies that could not be detected by human senses.

The third decade of the twentieth century, the period from 1917 to 1932 to be exact, brought an avalanche of new theories about atomic structure. Six years after his discovery of the atomic nucleus in 1911, Ernest Rutherford was working in his lab when he passed alpha particles though nitrogen gas and observed that some hydrogen was generated. He concluded that the alpha particles knocked out small bits from the nitrogen nucleus and reduced nitrogen atoms to hydrogen. He named those bits "protons."

This was not an ordinary chemical reaction. Transforming one element into another is no small feat. In fact, this was the first time such a transformation had been observed, and Rutherford became the first person to have caused a nuclear reaction. He theorized that protons had a positive electric charge and were densely concentrated in the nucleus. But the question was why did they not repel each other and blow away. What kept these positive charges so close to each other? There must be some other particles, electrically neutral, that create a strong nuclear force that keeps the protons densely packed in the nucleus. Rutherford thus predicted the existence of neutrons that would be discovered more than a decade later.

A groundbreaking contribution to quantum theory was made in 1924 by a French physicist named Louis de Broglie. A real duke by ancestry, de Broglie introduced his concept of electron waves while working on his doctoral thesis *The Theory of the Quanta*. The wave-particle controversy had troubled physicists for a long time and even more so after Planck's work on energy quanta, which had been endorsed by Einstein. The thesis examiners were unable to evaluate de Broglie's material and passed the thesis to Einstein for an assessment. Einstein agreed wholeheartedly with the theory, and de Broglie was awarded his doctorate.

De Broglie's fascinating idea was that, just as light can behave both like a wave and a particle, so can matter. Any particle of matter—an electron, an atom, a tennis ball—has a wavelength that is equal to Planck's constant divided by the particle's momentum. Essentially de Broglie was applying Planck's famous equation $E = h\nu$, not just to the photon but to all matter. De Broglie's theory of wave-particle duality was fully supported by Einstein and was experimentally verified in 1927 by the American physicists Clinton Davisson and Lester Germer.

So it turns out that all material objects display wavelike behavior. But how is it that we never see the tennis ball oscillate like a wave as it travels through the air? According to de Broglie's theory, the wavelength of the tennis ball oscillation is equal to Planck's constant divided by the momentum of the tennis ball. Planck's constant is a very small number; it is a decimal number with thirty-four zeros after the decimal point. The momentum of the tennis ball is a huge number because of its enormous mass, relatively speaking. The result of dividing a tiny number (Planck's constant) by a huge number (tennis ball momentum) is an extremely tiny number, and the wavelength of the tennis ball is therefore an extremely tiny number—a septillionth of a nanometer or less. This wavelength is so small compared to the size of the tennis ball that our human senses are unable to detect the oscillation. Remember, wavelength is the inverse of frequency. The oscillation of the tennis ball as it travels through air has such high frequency that it is beyond the sensing abilities of humans. Quantum events have such extremely short wavelengths (or high frequencies) that are way beyond the frequency response of human senses. When we observe macroscopic events of the human scale, our friend is Newtonian mechanics, not quantum mechanics.

De Broglie's hypothesis could explain why electrons behaved like particles in some experiments and like waves in other experiments. Like any other body, electrons could display both types of behavior. The idea of wave-particle duality could now resolve many outstanding issues and become established fact, opening up a whole new area of research in the new physics.

Two years after de Broglie published his hypothesis, an Austrian theoretical physicist named Erwin Schroedinger published a paper on wave mechanics and presented an equation that was to bring a new revolution in modern physics. A detailed explanation of Schroedinger's equation requires some rather complicated mathematics that is beyond the scope and character of this book. But a qualitative review is both possible and necessary. The Schroedinger equation tells us what we "can know" about the electron's position. More precisely, it tells us what we can know about the probability of finding the electron at a specific point in space.

In developing his equation, Schroedinger's starting point was the assumption that the wavelike behavior of the electron can only be predicted by a wavelike mathematical formulation. A reasonable assumption indeed, but an assumption that rests on what we want to prove!

Here is one of the most common forms of Schroedinger's equation:

$$\frac{\partial^2 \Psi}{\partial x^2} + \frac{8\pi^2 m}{h^2}(E - V)\,\Psi = 0$$

You will notice that we have only written the x-coordinate of the equation, as if electrons could only move on a straight line in one dimension, but we did this in order to simplify the equation. The y and z components have the same form, and all we need to do is substitute ∂x^2 by ∂y^2 or ∂z^2. If we write all three x, y, z terms, then we will have a complete description of the electron's position in three-dimensional space.

There are many possible ways to derive Schroedinger's equation. In 1926 German physicist Max Born presented a remarkably simple derivation.[20] He started with the wave equation from classical physics, a simple equation that describes the spatial distribution of the amplitude of a wave at a fixed point in time. He then substituted the

---

20 Jurgen Renn, *Schroedinger and the Genesis of Wave Mechanics*, Max Planck Institute website, URL = https://www.mpiwg-berlin.mpg.de/ Preprints/P437.PDF.

wavelength λ with h/p, which is the ratio of Planck's constant divided by the particle's momentum, as given by de Broglie's hypothesis. In one more substitution, he expressed the momentum as a function of kinetic energy, which is equal to total energy E minus potential energy V. That is why the term $(E - V)$ appears in the equation.

Schroedinger, a supreme mathematical logician, was able to combine in one equation classical wave mechanics, Planck's theory, de Broglie's theory, and the conservation of energy. Schroedinger's equation shows how a new mathematical formulation of known concepts can create new knowledge. In a letter to Schroedinger, Einstein writes: "The idea of your work testifies to genuine ingenuity!"

The protagonist in Schroedinger's equation is not a variable indicating the position, velocity, or energy of the particle but an abstract wave function Ψ. The best physical interpretation of Ψ is that it represents the probability of finding the particle at a certain position in space and time. So the particle is represented by a wave, which means that the particle is going to be an oscillator because it is assumed to be so. Its position will oscillate about the central point of its location. We note the unmistakable influence of de Broglie's hypothesis. But we have to clarify the fact that the particle will be found to oscillate only because Schroedinger has framed its position as a wave function, which was necessitated by some experimental results.

Schroedinger went backward. Starting from experimental results, which indicated wavelike behavior, he framed the particle in a mathematical wave equation. This is the very essence of inductive reasoning! Once the theory is developed, we must then use a deductive course of reasoning to confirm and validate it. The equation would have to confirm the known experimental results, but it would also, hopefully, explain results of future experiments. Otherwise the theory is self-fulfilling and cannot be confirmed.

Schroedinger's equation applies to any physical object, and we use both words "electron" and "particle" as the object of study in this discussion. We are very interested, of course, in what this behavior does to the electron's position in the atom. Schroedinger's equation does not show the electron to be moving in an orbit around the nucleus. In fact, it does not care if the electron is near or far away from a nucleus. The electron sits at a point in space and oscillates about that point. If we place it at an orbital level in Bohr's atom, it will oscillate about a point on the orbit, but nothing in Schroedinger's theory tells us that it will actually move along the orbit around the nucleus.

An incredible new theory was put forward in 1927 by German theoretical physicist Werner Heisenberg. Working as assistant to Niels Bohr in Copenhagen, Heisenberg was investigating the mathematical foundations of quantum mechanics and developed his uncertainty principle, a landmark theory that created a great deal of debate, controversy, and criticism—but became one of the cornerstones of modern physics nevertheless.

The uncertainty principle basically states that the position and the velocity of an object at a certain time cannot both be measured exactly and cannot both be calculated exactly, even in theory. The concept of exact position and exact velocity together at the same instant of time has no meaning in nature. Any attempt to accurately measure a particle's position will increase the error in the measurement of its velocity. The disturbance of the velocity is random and cannot be predicted or calculated. The more precisely the position is known, the less precisely the velocity is known and vice versa.

There are two statements in the above paragraph that are contradictory and were placed deliberately in order to highlight the interpretation problems of the theory. The statements are that (1) the concept of exact position and exact velocity together at the same instant of time has no meaning in nature and (2) any attempt to accurately measure a particle's position will increase the error in the measurement of its velocity.

We will say that the first statement is ontological as it deals with a description of nature irrespective of human presence. We will call the second statement epistemological as it deals with a human attempt to acquire knowledge. The critical question is: does the uncertainty exist independently of any measurement attempts, or does it arise from a disturbance caused by the measurement apparatus? For several years after Heisenberg's publication of the uncertainty principle, many scientists, Bohr and Einstein among them, made calculations and performed experiments to determine the root cause of the uncertainty, whether it was due to direct intervention or not.

The second statement is fairly easy to understand and agree with. Any attempt to measure a particle's position will change its velocity. In other words, we are allowed to think that the particle does have an exact position in space. We just cannot measure it with perfect accuracy. An attempt to detect a tiny particle and measure its position and velocity with our huge experimental apparatus will cause a change in the particle's velocity and therefore introduce an unpredictable random error in the position measurement. But our inability to measure it does not mean that an exact position does not exist.

So, we can live with the second statement, but the first statement is quite problematic. Let us restate it here: "The concept of exact position and exact velocity together at the same instant of time has no meaning in nature." This is difficult to understand and, once understood, difficult to agree with. It seems odd that the position of an object at an exact point in space can be described by a probability without the object ever physically occupying that exact point in space. It is a mathematical and logical oddity.

Probability functions exist in order to describe our best knowledge of phenomena. They are a property of our knowledge of phenomena, not a property of the phenomena themselves. Does this remind us of the Kantian dichotomy of noumenon versus phenomenon? It sure does! The human mind can only know the phenomenon; it is unable to penetrate the noumenon. We perceive a reality of the human perspective, not the absolute ontological reality.

Let us assume for a moment that we have the ability to visually observe the electron. In order to do that, a photon of light must collide with the electron and be bounced off it before reaching our eyes. By the time the photon reaches our eyes, the electron has already changed its position because of the collision. We have, therefore, a knowledge of the electron's position that is different from the position held just before the collision. This is the uncertainty that Heisenberg's principle describes.

Is this an ontological uncertainty or is it an uncertainty related to the observer's need to observe? We will opt for the latter idea, that Heisenberg's uncertainty is contextual and has no ontological implications. The electron has an exact position, which changes constantly when light falls on it. The uncertainty is observer related, just like relativity's time dilation and length contraction that we will discuss later on. Heisenberg himself proposed that a hypothetical gamma-ray microscope would provide a more accurate result of the electron's position because of the shorter wavelength of gamma rays. Heisenberg showed that the uncertainty in the measurement of the position of the electron is approximately equal to the wavelength of the gamma rays. This is quite an astounding statement, leading to the idea that the uncertainty is observer related.

We are not alone in having logical problems with the ontological interpretation of quantum mechanics and the uncertainty principle. Einstein famously said that "God doesn't play dice." Bohr responded in one of his papers "Einstein, don't tell God what to do."

Schroedinger, the de facto founder of the probabilistic view of the microcosm, was not very comfortable with the implications of the new quantum theory. In discussions about the probability interpretation of quantum mechanics, Schroedinger reportedly said, "I don't like it and I'm sorry I ever had anything to do with it." The quote cannot be traced back to a specific source but it is indicative of the interpretation problems of the theory. Schroedinger was convinced through his life that his equation is deterministic.

Speaking about the new physics, Heisenberg said in 1963 that what we observe is not nature itself but nature exposed to our method of questioning. So quantum theory does not really describe the system but rather the information that we can possibly have about the system based on our observation method. This admission by Heisenberg, made almost four decades after he published the uncertainty principle, should be as valid a rejection of the ontological interpretation as we can ever have.

Einstein was not living in 1963, and we will never know how he would have received Heisenberg's statement. Quite happily, we might guess. In all fairness, Heisenberg's idea applies to all science. What we observe is not the absolute ontological reality but rather a reality shaped by the human perspective, as we will discuss later on.

The uncertainty introduced by Schroedinger and Heisenberg in our understanding of the microcosm would have a permanent effect. Twentieth-century physics was put on a new path of indeterministic randomness. Nothing that is happening in physics today indicates that this path can be reversed back to the world of structure, determinism, and causality. Physicists had to learn a new mathematical formalism, a new language that could describe stochastics, randomness, and probabilities.

We should be careful not to confuse the concepts of causality and determinism. Causality and randomness are not incompatible. For example, if a free electron drifting aimlessly within a metal is hit by a beam of light, it will absorb a photon and start moving faster because of its increased energy. This is a causal outcome that began with the electron's random walk. Causality and randomness can coexist, and, in all likelihood, they do coexist quite a lot. A causal system is not necessarily a deterministic system.

You might ask the question, how can the microcosm have so much randomness while the macroscopic behavior is so orderly and predictable? When we throw a stone up in the air we can calculate its trajectory with precision by applying the laws of Newtonian mechanics. We can calculate how long the stone will be up in the air, where and how hard it will hit the ground. But we know that a chaotic world exists inside the stone's structure, a world of billions and billions of tiny particles aimlessly drifting in free space, colliding and exchanging energies with one another in a seemingly random fashion. The answer to the question seems to be that when large numbers of particles are involved, quantum effects average out. The laws of quantum mechanics asymptotically approach the laws of classical mechanics as the number of particles increases.

Copenhagen is not just the capital city of Denmark; it is also the hub of a very prominent interpretation of quantum physics. The Copenhagen interpretation was first proposed by Copenhagen physicist Niels Bohr and was embraced by his partner and collaborator Werner Heisenberg. It was supported by Max Born, Wolfgang Pauli, and others and opposed by Einstein, Schroedinger, de Broglie, Planck, Popper, and Bertrand Russell. It was popular during most of the twentieth century and is still rather widely accepted in the scientific community.

A complete, concise statement of the Copenhagen interpretation does not exist. The interpretation consists of a collection of ideas and principles espoused by a group of physicists and philosophers around Bohr and Heisenberg. A fundamental principle of the Copenhagen interpretation is that the new quantum physics must be described in ordinary language and not rely exclusively on mathematical symbolism.

The idea that the laws of quantum mechanics asymptotically approach the laws of classical mechanics as the number of particles increases is part of the Copenhagen interpretation and is known as the correspondence principle. Another idea is the complementarity principle, proposed by Niels Bohr. It says that the wave function expresses a necessary, fundamental wave-particle duality. This should be reflected in ordinary language descriptions of experimental results. Depending on the experiment, the wave particle can show wave or particle properties. The wave properties and the particle properties are two types of behavior of the same physical entity. The wave function is a complete description of a wave-particle. Any information that cannot be derived from the wave function does not exist.

Measurement processes have a special status in the Copenhagen Interpretation, but their role in producing peculiar effects is not clearly defined. It is agreed, however, that there are limitations in the conceptualization of interactions between microscopic entities on the one hand and the macroscopic apparatus on the other. The role of the observer is important in the Copenhagen interpretation of quantum theory, as it is in Einstein's special relativity theory that we shall be discussing later on.

In his book *Physics and Philosophy*, Heisenberg points out that there are symmetries in quantum theory, as in the symmetry between waves and particles and also between position and velocity. If we agree that these symmetries are a genuine feature of nature, then the Copenhagen interpretation cannot be avoided, according to Heisenberg.

Einstein and Schroedinger were among the most prominent dissenters of the Copenhagen interpretation. Einstein could never agree with the reliance on probabilities. He believed that physical reality exists independently of the observer, and the motions of particles are precisely determined. Einstein held that as long as it continues to rely on statistics, quantum theory is incomplete.

In spite of dissenters of the stature of Einstein and Schroedinger, the Copenhagen interpretation was embraced by most physicists at that time. The theory was so popular in the scientific community that many bright students came from as far away as America, Japan, and India to attend Heisenberg's lectures. The new generation of scientists spread the ideas of the Copenhagen doctrine around the world, especially in the 1930s and 1940s, when many scientists were dispersing away from Germany.

The debates for and against the Copenhagen theory were renewed in the last part of the twentieth century and are holding strong to this day. They cover a wide range of scientific, epistemological, and metaphysical issues. A theory that was supposed to interpret the quantum world has itself become the object of interpretation. The debates have been fascinating, to say the least, and have engaged the human intellect at its maximum possible level. This is the instant in history where the common ground between science and philosophy has been the greatest.

The lack of intuitiveness is pervasive in the new physics, especially quantum physics. The new theories defy our intuition about how the physical world is supposed to work. In our visible world, objects can't pass through each other, whereas in the quantum world, they pass through barriers that would appear impenetrable. We can imagine how difficult it was for physicists to make the giant leap from the classical worldview to quantum physics. But the lack of intuitiveness of the new science is due to lack of emancipation of the new theories. Emancipation of an idea means bringing the idea into the sphere of intuition and common sense.

We do not normally ask a question such as, "Why does light go through a glass window but not through a brick wall?" You will not ask a question like this unless you are a physicist! We all see in our every-day life, from the first hours of our life, that light goes through a glass window but not through a brick wall. This observation has become part of our intuition and common sense, and we no longer think about it. Infants know intuitively that objects exist even when they are not looking at them. Intuition and common sense are empirical, like all other knowledge. Intuition is not an inherent property of the mind but builds slowly with the emancipation of new experiences. As long as we are unable to directly experience phenomena of relativity and quantum physics, these theories will not be emancipated into intuition and will continue to remain outside of our sphere of common sense.

Summing up, we might say that the new physics of the micro-cosm, called quantum physics or quantum mechanics, rests on three pillar principles:

(1) Planck's theory that the smallest quantity of energy is the quantum (or photon) and the next one in size is two quanta.

Then we have three quanta, and so on. You cannot have one quantum and a half. That simply does not exist in nature. In Planck's theory we have the genesis of the physical energy entity.

(2) Schroedinger's theory that we can determine the probability of finding the electron somewhere in space, but that is about the extent of how much we can know,

(3) Heisenberg's theory that you cannot precisely know both the position and the velocity of an electron at the same instant of time.

It has been almost one hundred years since the beginning of quantum mechanics, and the theory still causes discussions and debates on interpretation. During a physics conference in 2013, a poll was conducted among participants who were, for the most part, physicists, mathematicians, and philosophers.[21] In a multiple-choice questionnaire, one of the most popular answers was "We'll have to wait and see." The question "What is your favorite interpretation?" received a remarkably even distribution of answers across seven different interpretations of quantum theory. Almost 50 percent of the participants answered that they had changed their favorite interpretation at least once. It seems that we are far from wide consensus on exactly what quantum theory tells us about nature, although there is hardly a serious physicist today who does not embrace quantum concepts.

Our ability to have a direct view of the microcosm is limited by the capabilities of our best microscopes, just as our ability to see the macrocosm is limited by our best telescopes. But this will soon change. The new STEHM microscope, built at the University of Victoria in Canada, will allow direct views at the atomic level for the first time ever. The Scanning Transmission Electron Holography Microscope (STEHM) will achieve a resolution approaching forty picometers.

---

21 C. Sommer, *Another Survey of Foundational Attitudes towards Quantum Mechanics*, URL = http://arxiv.org/pdf/1303.2719v1.

The picometer is a unit of length equal to one-trillionth of a meter. The length of the picometer is such that its use is entirely confined to particle physics and quantum physics. Atoms are between sixty and five hundred picometers in diameter, and that puts the STEHM viewing field right inside of the atom. As amazing as this sounds, it is happening in our day! The microscope will provide scientists with three-dimensional information at the atomic level and will push the boundaries of research in many areas, including quantum physics, chemistry, materials science, nanotechnology, biology, and medicine.

The avalanche of new discoveries in the late nineteenth century mounted the pressure under the volcano of scientific knowledge, and an eruption was in the making. The eruption occurred in the first three decades of the twentieth century, with the relativity theory and quantum theory creating a new concept of our macrocosm and microcosm. The road to our present understanding of natural phenomena was a long road with different theories and models along the way. Each model made a rather smooth transition from previous theories. Each model had elements from past models. We might say that each model was a synthesis of elements of past models, with some new elements. We might also say that the course of development of our present knowledge has been evolutionary rather than revolutionary. This observation will be useful to us later when we discuss Thomas Kuhn's theory on the structure of scientific revolutions.

Positively Logical

The New Analytic Philosophy

THE TWENTIETH CENTURY opened with a bang. Max Planck's quantum theory and Albert Einstein's relativity theory were the Big Bang in physics and created an incredible chain of scientific activity in new areas of exploration. Igor Stravinsky's *Rite of Spring* in 1913 was literally a bang, as its first performance in Paris started a riot. It was a revolutionary work that pushed the boundaries of musical design into unforeseen new areas and had a profound influence through the entire century. Pablo Picasso's and George Braque's paintings were the Big Bang in art, taking apart and analyzing objects in terms of their shapes and creating a new avant-garde of lasting influence. The Big Bang in world developments came with the break of World War I in 1914—a war that killed tens of millions of combatants and civilians and led directly to World War II, the most devastating war in history.

In philosophy the bang was not big, but a school with a brand-new orientation developed nevertheless. It became known as analytic philosophy and focused on clarity and argument via a new logical, linguistic, and mathematical formulation of philosophical ideas. Although this new school of analytic philosophy dominated the philosophy of

English-speaking countries, its forerunner was an Austrian physicist and philosopher named Ernst Mach, and its founder was a German mathematician and philosopher named Gottlob Frege.

Ernst Mach made important contributions in both physics and philosophy and is best known for the Mach principle, a term coined by Albert Einstein. Mach was born in 1838 in Brno, Austria, which is now a city of the Czech Republic. He studied physics in Vienna and became professor of mathematics in Graz. Mach did extensive work in physics, mainly in the Doppler effect, optics, wave dynamics, and aerodynamics. The Mach principle is the idea that local inertial frames are determined by the large-scale distribution of matter. A more generalized statement of the principle is that local physical laws are determined by the large-scale structure of the universe. Mach's principle is suggestive of a connection between matter, and geometry, and it was an important factor that influenced Einstein's general relativity theory. Einstein attempted to formalize Mach's principle with a mathematical expression, but his effort was not successful. There is a tremendous body of scientific and philosophical literature with various interpretations of Mach's principle and specific efforts to find mathematical connections between the principle and Einstein's general relativity.

Mach developed a philosophy of science that saw scientific laws as mathematical expressions of sensations rather than reality as it may exist beyond sensations. In Mach's own words: "In reality, the law always contains less than the fact itself, because it does not reproduce the fact as a whole but only in that aspect of it which is important for us, the rest being intentionally or from necessity omitted."[22] Mach's idea is very similar to the theory of knowledge of his contemporary Charles Peirce.

Gottlob Frege is recognized as the founder of twentieth-century analytic philosophy. He was born in 1848 in Wismar, a seaport

---

22 Ernst Mach, "The Economical Nature of Physical Inquiry", *Popular Scientific Lectures* (1898): 192.

city near Hamburg, and studied mathematics and physics at the Universities of Jena and Goettingen. Following graduation, his interests turned to logic; and in 1879 he wrote his major work, titled *A Formal Language for Pure Thought Modeled on That of Arithmetic.* Frege formalized the notion of proof in a logical system that is still accepted today. He developed a comprehensive system relating and merging logic, mathematics, and language that continues to provide valuable insights to philosophers. Frege's work laid the foundation for a whole new system of analytic philosophy that would bring linguistic concepts and mathematical symbolism into philosophy and would begin to blur the distinction between philosophy and science.

Movies about philosophers are not common, but Austrian philosopher Ludwig Wittgenstein was the subject of a 1993 film, simply titled *Wittgenstein.* A film about thinking is not something we see every day, but Wittgenstein deserves the distinction as much as any modern philosopher. Considered by many people as the twentieth century's most important philosopher, Wittgenstein was born in Vienna in 1889 into one of Europe's wealthiest families. Both of Ludwig's parents were of Jewish ancestry, but all children were baptized as Catholics. Karl Wittgenstein, Ludwig's father, was a tycoon who controlled an effective monopoly of iron and steel within the Austro-Hungarian empire. A highly educated and cultured man, his patronage of the arts was often compared with Andrew Carnegie's. The Wittgenstein house in Vienna was a spectacular palace and a cultural meeting place for the intellectual elite. Sigmund Freud was a regular visitor, and composers Johannes Brahms and Gustav Mahler would often perform there. In Bruno Walter's words, the Wittgenstein palace had an "all-pervading atmosphere of humanity and culture."

Ludwig was taught by private tutors at home until the age of fourteen. Lacking formal schooling, Ludwig failed his entrance exams in the schools of choice and ended up in a small technical school, where he completed his secondary education. After graduation he studied engineering in Berlin, but he gradually became interested in mathematics—an interest that turned to obsession, according to his sister. In 1911 he visited Frege at the University of Jena and was advised by Frege to study with Bertrand Russell, who taught at Cambridge. Wittgenstein studied under Russell at Cambridge, and this was the beginning of a long friendship of mutual admiration and intellectual rivalry at the same time.

It did not take long for Russell to realize that Wittgenstein was a genius. Russell once confided to his lover, the famous aristocrat Lady Ottoline Morrell, that Wittgenstein would solve the problems that he (Russell) was too old to solve. Wittgenstein criticized Russell's work in 1916, and Russell confessed in his autobiography that Wittgenstein's criticism was an event of great importance in his life and had affected everything he had done since. Russell saw that Wittgenstein was right and that he (Russell) could not ever again hope to do any fundamental work in philosophy.

In 1913 Wittgenstein decided to get away from academic circles and retreated to a small village in Norway, where he could collect his thoughts. This was when he drafted *Tractatus*, his only published book. He was now a wealthy man, having inherited his father's fortune. Wittgenstein donated some of his money to artists and writers, and after the war, he gave his entire fortune to his brothers and sisters.

There is so much more to this man's incredible life. When World War I broke out, he volunteered for military service, even though he could be exempted for medical reasons. He spent the first two years of the war behind the lines, relatively safe and able to continue his work on logic. In 1916, however, at his own request, he was sent to a fighting unit at the Russian front and was decorated several times for bravery.

Wittgenstein published *Tractatus* in 1922. Believing he had answered all the essential problems of philosophy, he retreated to a small village in Austria where he taught elementary school until 1929, when he returned to Cambridge as professor of philosophy. At this stage he was abandoning some of the logical forms as they appeared in *Tractatus*. His later writings were collected and published after his death as *Philosophical Investigations*. This is regarded as his most mature and important work and is the most authoritative source of Wittgenstein's ideas.

In his introduction to *Tractatus,* Bertrand Russell writes: "Mr Wittgenstein maintains that everything properly philosophical belongs to what can only be shown, to what is in common between a fact and its logical picture. It results from this view that nothing correct can be said in philosophy. Every philosophical proposition is bad grammar, and the best that we can hope to achieve by philosophical discussion is to lead people to see that philosophical discussion is a mistake."[23]

Wittgenstein attacked the very nature of philosophy. He believed that what we find in philosophy is trivial; it does not teach us new facts—only science does that. But the proper synopsis of these trivialities is enormously difficult and has immense importance. Philosophy is in fact the synopsis of trivialities. Philosophy creates problems out of an inflexible use of language. For example, we use the word "mind" in our everyday language without any difficulty—until a philosopher asks "What is the mind?" We then imagine that this question has to be answered by identifying some "thing" that is the mind. If we remind ourselves that language has many uses and that words can be used quite meaningfully without corresponding to things, the problem disappears.[24]

What is required, according to Wittgenstein, is not a correct doctrine but a clear view, one that dispels the confusion that gives rise to the problem. We often mistake grammatical rules for material propositions, and that adds to the philosophical confusion. For example, the expression 2 + 3 = 5 is a rule of grammar, according to Wittgenstein,

23 Ludwig Wittgenstein, *Tractatus Logico-Philosophicus*, Project Gutenberg website, URL = http://www.gutenberg.org/files/5740/5740-pdf.pdf

24 *Encyclopedia Britannica*, http://www.britannica.com/EBchecked/topic/646252/Ludwig-Wittgenstein.

not a proposition that describes reality. It would be interesting to note here that philosopher A. J. Ayer later claimed that all mathematical expressions are tautologies—that is, they tell us nothing other than A = A. We will review this idea when we discuss Ayer later in this chapter, but we already see the seeds of Ayer's idea here in Wittgenstein.

Wittgenstein believed that language derives its meaning from the way that it is used and from context. Let us look at a simple example. The statement "Person A is great" can be used as "Beethoven is great" and also "My friend John is great." Those around me will know immediately that my intention as to the scale of greatness is quite different in my two sentences. That is what Wittgenstein refers to as context. The question is, how do we modify the form "Person A has quality B" to account for this difference in context?

Context adds flexibility to language. Context allows the mapping of a flexible set of linguistic variants into an idea. A person working for pleasure experiences different psychological consequences compared to a person working for survival. The concept of work is different for the two persons because the context is different. Context adds different meaning to the same words and ideas, and its role has been overlooked in philosophy.

In the preface to *Tractatus,* Wittgenstein says, "What can be said at all can be said clearly, and what we cannot talk about we must pass over in silence."[25] It is obvious that Wittgenstein regards language as a central aspect of knowledge. Wittgenstein was an exceptionally intellectual thinker and was certainly not the scholar stereotype. His work has been tremendously influential in the twentieth century. His great contribution to philosophy is that he opened up a whole new area of exploration, the philosophy of language.

The new linguistic focus would rail philosophy into new tracks and would distort and blur the boundaries between philosophy, logic,

---

25 Ludwig Wittgenstein, *Tractatus Logico-Philosophicus,* Project Gutenberg website, URL = http://www.gutenberg.org/files/5740/5740-pdf.pdf

mathematics and science. Bertrand Russell summarizes philosophy's new direction as follows:

> Modern analytical empiricism, of which I have been giving an outline, differs from that of Locke, Berkeley, and Hume by its incorporation of mathematics and its development of a powerful logical technique. It is thus able, in regard to certain problems, to achieve definite answers, which have the quality of science rather than of philosophy. It has the advantage, in comparison with the philosophies of the system-builders, of being able to tackle its problems one at a time, instead of having to invent at one stroke a block theory of the whole universe. Its methods, in this respect, resemble those of science. I have no doubt that, in so far as philosophical knowledge is possible, it is by such methods that it must be sought; I have also no doubt that, by these methods, many ancient problems are completely soluble.[26]

Russell is one of the architects of the new philosophy, together with Frege, Wittgenstein, Whitehead, and Moore. He enjoyed a long, remarkable life of ninety-eight years, spanning the last third of the nineteenth century and the first two-thirds of the twentieth. He is best remembered for his contributions to logic, mathematics, and philosophy, but he was also an important historian, social critic, and political activist. The reader is familiar by now with some of Russell's ideas as we have often included his thought-provoking commentary on various theories in this book.

Russell's 1945 book *History of Western Philosophy* is his best-known work. It is a masterful treatment and a clearly written historical review of philosophers and their theories, going back to the seventh century BC. Russell was a great educator and did much to inform the general

---

26 Bertrand Russell, *The History of Western Philosophy* (New York: Simon & Shuster, 1945), 834.

public with popularizing books, such as *The ABC of Relativity* and *The Problems of Philosophy*. Russell's main philosophical work is the massive three-volume *Principia Mathematica*, written in 1910–113 and coauthored with English mathematician and philosopher Alfred North Whitehead. The book is considered one of the twentieth century's most important works on mathematical logic. According to the *Encyclopedia Britannica*, Russell's and Whitehead's "theory of types," as formulated in the book, is so bewilderingly complex that very few people, whether philosophers or mathematicians, have made the gargantuan effort required to master the details of this monumental work.[27]

We will avoid the temptation to join that select group of people for a very good reason: Russell's theory has much more to do with the mathematical and logical formalism than with the nature of scientific knowledge, which is our primary interest in this book. Russell essentially was attempting to prove that mathematics is a branch of logic and can be reduced to a system of logical formalism. The philosophical significance of the *Principia Mathematica* is still a matter of debate.

Kurt Gödel, one of the most important mathematicians of the century, provided a proof that there cannot be a system of logic from which the whole of mathematics can be derived. According to Gödel's theorem, in all logical systems, there are statements, expressed in the system's own language, which you can neither prove nor refute. You can try to settle such an undecidable statement by adding whatever axioms are necessary to prove it, but according to Gödel's result, other undecidable statements will pop up elsewhere. In other words, each logical system contains more true statements that it can possibly prove by using its own rules. There are, therefore, limits of provability in all axiomatic systems. Gödel's idea has many important consequences on several different areas, one of which is that mathematics is not reducible to a system of logic, as Russell had tried to prove. In any event,

27 http://www.britannica.com/EBchecked/topic/513124/Bertrand-Russell.

most philosophers today recognize that the influence of Bertrand Russell's work on the development of mathematical logic through the twentieth century has been immense.

Frege, Wittgenstein, and Russell opened up new paths of philosophical inquiry that led to a new school of thought called analytic philosophy. One of the most important thinkers along this new path was an Austrian named Karl Popper. He was born, raised, and educated in Vienna. His parents were upper-middle-class people of Jewish background who had converted to Lutheranism in order to assimilate culturally in Austrian society. Karl's father was a prominent lawyer and bibliophile with a personal library of thousands of volumes. Raised in this intellectual environment, Karl took an interest in classics, philosophy, and music and acquired from his father an interest in social and political issues.

As a student at the University of Vienna, Popper became involved in Marxist ideology and activism but soon got disillusioned with pseudoscientific dogmatic thinking and became a supporter of social liberalism, which, in those days, meant individual liberty; social justice; market economy; and a big role of government in addressing social issues such as poverty, health, and education. During his student days and after graduation, Popper worked in various jobs as construction worker, cabinetmaker, volunteer in children's hospitals, schoolteacher, and finally university lecturer in New Zealand. In the meantime he wrote his first book in 1934, titled *Logic of Scientific Discovery*.

After the war, Popper moved back to Europe and took academic positions teaching logic and scientific method. Popper discovered the psychoanalytic theories of Freud and Adler and was fascinated by a lecture that Einstein gave in Vienna on relativity theory. The dominance of the critical spirit in Einstein and its total absence in Marx, Freud, and Adler struck Popper as being of fundamental importance: the pioneers of psychoanalysis, he came to think, couched their theories in terms that made them amenable only to confirmation; however, Einstein's theory, crucially, had testable implications, which, if false, would have falsified the entire theory. Popper thought that Einstein's theory, as a theory properly grounded in scientific thought and method, was highly risky, in the sense that it was possible to deduce consequences from it that would undermine the whole theory if turned out to be false. In contrast, nothing could falsify psychoanalytic theories. He thus came to the conclusion that such theories had more in common with primitive myths than with genuine science. This led Popper to conclude that what was perceived as the remarkable strengths of psychoanalytical theories were actually their weaknesses. Psychoanalytical theories were crafted in a way that made them able to refute any criticism and give an explanation for every possible form of human behavior. The nature of such theories made it impossible for any criticism or experiment to show them to be false.[28]

This concept of falsifiability, first introduced by Popper, has since become an important criterion in the philosophy of science and will be a recurring theme in the rest of this book. A central property of science is that every genuine scientific claim is capable of being proven false. The strength of a scientific theory lies in its both being susceptible to falsification and not actually being falsified. Popper considered that if a theory cannot be falsified by criticism, it is not a scientific theory. The term "falsifiable" does not mean that something is proven false. It means that, if something is false, its falsity can be proven by observation or experiment. Popper stresses the problem of demarcation, distinguishing the scientific from the unscientific, and makes falsifiability the criterion of demarcation. For a theory to be scientific, it must be falsifiable.

Popper was not at all in agreement with the new wave in early twentieth-century philosophy who claimed that linguistic difficulties rendered philosophy useless. In Popper's own words:

---

28 *Stanford Encyclopedia of Philosophy*, http://plato.stanford.edu/entries/ popper/.

Language analysts believe that there are no genuine philosophical problems, or that the problems of philosophy, if any, are problems of linguistic usage, or of the meaning of words. I, however, believe that there is at least one philosophical problem in which all thinking men are interested. It is the problem of cosmology: the problem of understanding the world, including ourselves and our knowledge, as part of the world. All science is cosmology, I believe, and for me the interest of philosophy, no less than of science, lies solely in the contributions which it has made to it. For me, at any rate, both philosophy and science would lose all their attraction if they were to give up that pursuit. Admittedly, understanding the functions of our language is an important part of it; but explaining away our problems as merely linguistic puzzles is not.[29]

We may recall from our discussion of David Hume how he objected to the method of inductive reasoning. Hume argued that drawing inferences from the observed to the unobserved results in circular reasoning as it invokes in the proof the principle of uniformity, which requires induction to prove its own validity. Popper is equally opposed to induction. The test of falsifiability requires deductive reasoning, starting from the theory or hypothesis and going down to experimental observations designed to falsify the hypothesis.

Let us reflect for a moment on what inductive reasoning really means for us nonphilosophers and what our attitude might be with respect to Hume's and Popper's objections. The crux of the objections is that, when trying to prove that a hypothesis is true, our observations cover only a very small subset of all possible empirical instances of the hypothesis. So, we prove the hypothesis for a small subset of observed instances, and then we infer that the hypothesis is true for

---

29 Karl Popper, *Preface to "The Logic of Scientific Discovery*," 1st English ed., (New York: Routledge, 1959), XVIII.

all possible unobserved instances. We may infer that the conclusion drawn is not an absolute, fixed, and final truth. It would take observations of an infinite set of instances to demonstrate the absolute truth of our hypothesis. However, it would take a single observation to disprove the hypothesis via deductive reasoning.

This argument is a heuristic proof that Hume's and Popper's objections to inductive reasoning are valid. If, however, we accept that there are no fixed and final truths, then Hume's and Popper's objections must disappear until a method of reasoning is invented that can produce such absolute, fixed, and final truths.

The fact is that scientists do not really care if they use inductive, deductive, or some other type of reasoning. In most cases they will use both inductive and deductive reasoning and will try everything else to prove or disprove a theory. We also must consider that "proof" in science is something less than "proof" in the Euclidean sense. In addition to inductive and deductive reasoning, a scientific hypothesis must pass the scrutiny of the scientific community, which is an important element of the validation of the theory; and even then, we may not have a final scientific truth.

The works of Frege, Wittgenstein, Russell, and Popper led to the creation of a new school of philosophy that emphasized language, logic, and verification. Various fancy names were given to this new school of thought, but the one that prevailed was logical positivism. In the new doctrine, scientific knowledge is the only kind of factual knowledge, and all traditional metaphysical doctrines are to be rejected as meaningless.

It seemed that this new philosophy was offering a powerful new vision and had everything going for it except for a fervent popularizer—until Alfred Jules Ayer appeared on the scene. The basic premise of the new philosophy was summed up by Ayer in very simple terms: "We say that a sentence is factually significant to us if and only if we know how to verify the proposition which it purports to express, that is, if we know what observations would lead us, under certain conditions, to accept the proposition as being true or reject it as being false."[30]

In other words, Ayer says that something has meaning only if it can be tested to be true or false. In essence, logical positivism leads all knowledge to logical, scientific foundations. In endorsing these views, Ayer saw himself as continuing along the British empiricism of Locke and Hume and the most recent representative of that empiricism, Bertrand Russell.

Ayer was born in 1910 in London to a wealthy family. He excelled in classics at a young age but had no opportunity to study science, an omission he would always regret. He studied philosophy at Oxford on a scholarship. Ayer served in World War II, spying for the Secret Intelligence Service. Well connected, extroverted, a socializer, bon vivant, and a sports fan, Ayer was not exactly an academic stereotype. He married four times, including his marriage to Vanessa Salmon, the mother of famous TV celebrity cook Nigella Lawson.

Ayer published his first book in 1936 under the title *Language, Truth and Logic*. The book was successful immediately. Ayer's crisp writing and clear logical sequences make the book excellent philosophical reading. Ayer says that the views in *Language, Truth and Logic* derive from the doctrines of Bertrand Russell and Wittgenstein, which are themselves the logical outcome of the empiricism of Berkeley and Hume.

To test whether a sentence expresses a genuine empirical hypothesis, Ayer adopts what he calls a verification principle. He says:

> For I require of an empirical hypothesis, not indeed that it should be conclusively verifiable, but that some possible sense experience should be relevant to the determination of its truth or falsehood. If a putative proposition fails to satisfy this principle, and is not a tautology, then I hold that it is

---

30 A. J. Ayer, *Language, Truth and Logic*, (London: Victor Gollancz, 1936), 35.

metaphysical, and that, being metaphysical, it is neither true nor false but literally senseless. It will be found that much of what ordinarily passes for philosophy is metaphysical according-ing to this criterion, and, in particular, that it can not be significantly asserted that there is a non-empirical world of values, or that men have immortal souls, or that there is a transcendent God.[31]

It is rather obvious that Ayer does not have much use for meta-physics. He thinks much or all of it meaningless, as none of its theo-ries can be verified. Ayer frequently attacked metaphysics and was a strong critic of the German philosopher Martin Heidegger's vast, all-embracing theories about existence. These, he felt, were completely unverifiable through empirical demonstration and logical analysis. This useless philosophy was an unfortunate strain in modern thought, and Heidegger, according to Ayer, was its worst example.

Statements of fact or value have meaning only insofar as they are verifiable. Even ethical and aesthetic judgments are meaningless unless they can be subjected to empirical testing. An empirical test may be practical or theoretical. The a priori statements of logic and mathematics do not claim to provide factual content. Those state-ments can be said to be true only because of the conventions that gov-ern the use of the symbols that make up the statements. Ayer accepts the existence and need for a priori knowledge of necessary proposi-tions but rejects Kant's idea that any of these necessary propositions are synthetic. They are, without exception, analytic propositions, or tautologies. These propositions do not make any assertion about the empirical world and cannot be refuted by experience. They simply record our determination to use words in a certain fashion. According to Ayer, all mathematics consists of tautologies and cannot contain any empirical knowledge derived from the natural world.

31 A. J. Ayer, Preface in *Language, Truth and Logic*, (London: Victor Gollancz, 1936), 2.

We need to be careful to interpret the last statement correctly. Ayer is referring to pure mathematical forms that are devoid of empirical content. We can understand this with an example. Let us consider the mathematical expression $x = ya^2$, where x and y are variables and a is a constant. This expression has no physical equivalent in the natural world; it is a pure mathematical expression relating the magnitudes of two abstract variables. This is an analytic expression that is absolutely and a priori true; it has no need for verification. In fact, it cannot be verified by any method. It is definitional; it is true under all conditions; it is a tautology like $A = A$. If, however, we assign physical properties to x, y, and a and assert that the physical properties are related in the same way $x = ya^2$, then we have an empirical import, and the mathematical relation is no longer analytic, it is synthetic. For example, if we assign energy to x, mass to y, and the speed of light c to a, our relation becomes $E = mc^2$, the famous Einstein equation. But now this relation is no longer self-evident; it requires evidence and proof. The correct interpretation of Ayer's assertion is that $x = ya^2$ is a tautology, but $E = mc^2$ is not. The former is an abstract form—absolutely true, unverifiable, and contains no knowledge. The latter is an empirical form, must be verified, and, if it passes the test, contains new knowledge.

Ayer further believes that much of philosophy is analytic and therefore has no meaning of its own. Other parts of philosophy, such as metaphysics, theology, ethics, and aesthetics, focus on ideas that cannot be judged as being true or false, and it is therefore meaningless to discuss them. Ethical and aesthetic statements, in particular, are expressions of feelings, have no factual content that can be empirically tested, and cannot therefore be argued for or against.

Ayer does not see philosophy as a metaphysical concern that attempts to speculate on unsolvable ontological problems. Instead, philosophy is an activity of defining and clarifying the logical relationships of empirical propositions. Ayer's work restricts the role of philosophy in metaphysical areas and expands it in other unforeseen areas, such as artificial intelligence. In Ayer's words:

> The distinction between a conscious man and an unconscious machine resolves itself into a distinction between different types of perceptible behaviour. The only ground I can have for asserting that an object which appears to be a conscious being is not really a conscious being, but only a dummy or a machine, is that it fails to satisfy one of the empirical tests by which the presence or absence of consciousness is determined.[32]

In this excerpt, Ayer is considering the distinction between a conscious man and an unconscious machine and anticipates the 1950 development of the Turing test to determine a machine's capability of humanlike intelligence. We will discuss the Turing test later along with other related issues of artificial intelligence.

The ideas of Wittgenstein, Russell, Popper, and Ayer were rooted in the philosophies of the British empiricists Locke, Berkeley, and Hume and became collectively known as logical positivism. They dominated twentieth-century philosophy and are still the basis of much philosophical discussion today. Their simple premise, that any theory cannot create new knowledge unless validated by empirical facts, changed the nature of philosophy. This is empiricism at its very best, an empiricism based on solid scientific principles.

---

32 Ayer, *Language, Truth and Logic*, 86.

Einstein's Universe

Relativity in Space-Time

W HEN EINSTEIN FIRST published his theory of relativity in 1905, it was said that there were no more than a few scientists in the world who could understand it. Sir Arthur Eddington, an English astrophysicist and one of the early champions of relativity, was asked whether it was true that he was one of only three people in the world who understood relativity. He replied, "Who is the third?" He implied, obviously, that he and Einstein were the only two people who understood the theory, and there was no third person.

In this chapter we will make a good effort to become that third person—although the number today, more than one hundred years later, is much larger. We will make the effort, with and without the math, as it is not possible to understand the revolution of modern physics without a good appreciation of basic relativity.

Albert Einstein was a true intellectual giant, and his name today is synonymous with genius. He was born in Ulm on March 14, 1879. Built on the banks of the Danube river, Ulm was best known for having the tallest church in the world. Today it is also known as the birthplace of Albert Einstein. He was born into a middle-class, nonobservant Ashkenazi Jewish family. The Ashkenazi are a Jewish ethnic division with roots in the Holy Roman Empire. They established communities throughout central and eastern Europe and made remarkable contributions to philosophy, literature, arts, music, and science. Some of the great people who came from Ashkenazi communities are Sigmund Freud, Felix Mendelssohn, Marc Chagall, and, of course, Albert Einstein. Six weeks after Albert was born, the family moved to Munich, where Albert's father ran an electrochemical factory. Albert did not speak until he was three years old, and his performance in elementary school was rather mediocre. Einstein has written about two events that had a marked effect in his early childhood. When he was five years old, his father gave him a pocket compass. The young Albert was fascinated with the fact that something was causing the compass needle to move. When Albert was ten years old, his family invited Max Talmud, a poor medical student from Poland, to come for dinner to the Einstein household and tutor young Albert, introducing him to mathematics and philosophy. Talmud gave Albert Immanuel Kant's *Critique of Pure Reason* as well as Euclid's *Elements*, a book that Einstein read over and over again and called the "holy little geometry book."

Young Albert was a little rebel at school, often protesting about excessive memorization and repetition and the lack of creative learning. Einstein knew at such young age that there is a big difference between memorizing formulas and understanding concepts. In mathematics, physics, and philosophy, Einstein studied from extracurricular books and was well ahead of his classmates and the school curriculum.

Under his mother's influence, the young Albert developed a love for classical music and started the study of the violin at the age of five. At the age of thirteen, he discovered the violin sonatas of Mozart and later the piano sonatas of Beethoven. Music became a very important aspect of his life, and he often played in chamber groups together with professional musicians. Later, in his adult life, Einstein would see the purity and beauty of Mozart's music as a reflection of the inner beauty of the universe.

When Albert was fifteen years old, his father's company failed, and the family moved to Italy but Albert stayed in Munich to complete his high-school education. One year later Albert failed the entrance exams for the Swiss Federal Polytechnic in Zurich and was urged to complete his secondary schooling in Switzerland, which he did. He was then admitted at the Zurich Polytechnic at the age of seventeen in a program that would train him as a teacher of physics and mathematics. In the meantime he had renounced his German citizenship in order to avoid military service. He received his diploma in 1901 and acquired Swiss citizenship in the same year. After two frustrating years searching for a teaching job, he was employed as an assistant examiner in the Swiss Patent Office in Bern.

Einstein's work in the evaluation of patent applications was not as dull as we read sometimes in Einstein biographies. Much of the work was quite challenging, from a scientific point of view, and related to questions about the transmission of electrical signals and electro-mechanical synchronization of time. One patent Einstein was faced with dealt with how to synchronize clocks across the vast network of European train lines and stations so that trains traveling in opposite directions on the same track did not collide. Such challenges shaped Einstein's thought experiments, which would eventually culminate in his radical theories about the nature of light and the fundamental connection between space and time.

In 1902 Einstein started the Olympia Academy, a small group of friends and colleagues who met in Einstein's apartment in Bern to discuss science and philosophy. Their meetings would often last until the early morning hours, and the discussions played a significant role in Einstein's intellectual development. The reading suggested by Einstein for the group's first meeting was *The Grammar of Science*, a book by English mathematician Karl Pearson, whose ideas revolved around themes encountered later in Einstein's scientific work—ideas such as an observer's perception of the laws of nature, an observer's contraction of time when he travels near the speed of light, and other relativistic ideas. The effects of these ideas on Einstein will be better understood when we explore the meaning of Einstein's theories. In 1905 Einstein obtained his doctoral degree at the University of Zurich. He was now married to Mileva Maric, his sweetheart from the Zurich Polytechnic.

This same year, 1905, turned out to be a landmark year for Einstein and for all science. Einstein wrote four groundbreaking papers on the photoelectric effect, Brownian motion, special relativity, and the equivalence of mass and energy. The papers were initially ignored by the scientific community, until they received the attention of Max Planck, the most influential scientist at that time. With Planck's endorsement, Einstein rose rapidly in the scientific community; and by 1908 he was recognized as a leading scientist. A series of academic appointments followed at universities in Bern, Zurich, and Prague, leading to a full professorship at the University of Berlin in 1914.

One year later Einstein completed his general theory of relativity, which had occupied him for years. The theory predicted that light passing by a large mass, such as the sun, would be deflected by gravity. The theory is a gravitational theory and describes relationships between space, time, mass, and gravity. It was published in 1915; and four years later, the English physicist Arthur Eddington organized an expedition to an island off the west coast of Africa to observe a solar eclipse. Eddington's observations showed that distant stars that were behind the sun, as seen from the earth, were actually visible. This could not be seen except during a solar ellipse. The moon comes between the sun and the earth, blocks the sun, and creates the necessary darkness to make the stars visible. Eddington's star experiment confirmed that light from the stars was deflected by the sun's gravity, displacing the star image and making the star visible. Einstein's theory that light is deflected by gravity was confirmed. A theory invented by a German was confirmed by an Englishman from Newton's university! And all this one year after World War I, in which Germany and Britain had fought each other to total destruction. It was a great subject for newspaper headlines and, for many of us, an uplifting picture after the traumatic war experience.

Following the experimental confirmation of his theory, the outpouring of international acclaim brought worldwide fame to Einstein. In 1921 he received the Nobel prize for physics, not for the relativity theory, which was still controversial, but for his explanation of the photoelectric effect. In the same year Einstein made his first trip to the United States, a country that was becoming a haven destination for European scientists fleeing authoritarian regimes. It was rather suddenly becoming the preeminent center of pioneering research in science.

Einstein delivered several lectures at Columbia University and Princeton University. He was quite impressed with the scientific achievements already made there, with and without the science refugees. He was also impressed with the cultural aspects of everyday life, saying that "what strikes a visitor is the joyous, positive attitude to life. The American is friendly, self-confident, optimistic and without envy."[33]

Einstein was a pacifist and said that science was often inclined to do more harm than good. His aversion to war led him to befriend other pacifists, like author Upton Sinclair and film star Charlie Chaplin. His first meeting with Chaplin during a tour of Universal Studios in Los Angeles was a memorable experience for both men. They had instant rapport and Chaplin invited Einstein and his wife Elsa (Einstein was now in his second marriage) to his home for dinner. Chaplin said that Einstein's calm, gentle persona seemed to conceal a highly emotional

---

33 Gerald Holton, *The Advancement of Science and its Burdens*, (Cambridge, Mass: Harvard University Press, 1998), 127.

temperament, from which came his extraordinary intellectual energy. Chaplin later recalled Elsa telling him about the time Einstein conceived his theory of relativity. During breakfast one morning, Einstein seemed lost in thought and ignored his food. Elsa asked him if something was bothering him. He sat down at his piano and started playing. He continued playing and writing notes for half an hour and then went upstairs to his study, where he remained for two weeks, with Elsa bringing up his food. At the end of the two weeks, he came downstairs with two sheets of paper bearing his theory.[34]

In the 1930s, during the rise of the Nazis in Germany, Einstein's theories on relativity became a target of the new regime. In 1931 the Nazis enlisted other physicists to denounce Einstein and his theories as "Jewish physics." At this time the new German government had passed a law barring Jews from holding any official position, including teaching at universities. Einstein learned that his name was on a list of assassination targets, and a Nazi organization published a magazine with Einstein's picture and the caption "Not Yet Hanged" on the cover, offering a $5,000 reward on his head.

Einstein's departure from his native country was now a matter of time, and he managed to leave Germany in 1932. One year later he renounced his German citizenship and turned his passport to the German consulate in Antwerp, Belgium. Nazi Propaganda Minister Joseph Goebbels singled out Einstein, a Nobel laureate, as the prime target in his war against Jewish intellectualism. He ordered that Einstein's belongings be confiscated, his books burned, and his cottage turned into an Aryan training camp. Goebbels declared that "Jewish intellectualism is dead." Einstein's reaction to the burnings was rather stoic. He understood that fascism feared the influence of men of intellectual independence.

In 1933 Einstein went back to the United States and took a position at the Institute for Advanced Study at Princeton, New

---

34 Charles Chaplin, *My Autobiography* (New York: Simon and Schuster, 1964), 320.

Jersey. It was here that he would spend the rest of his career trying to develop a unified field theory, a general theory that would unify the forces of the universe and the laws of physics into one framework. Einstein had offers during this time to work in European universities, including Oxford, but decided to stay at Princeton and apply for American citizenship. He had found happiness in his new home in friendly America and in the liberal atmosphere at Princeton.

In 1939, a group of scientists, concerned that Germany was doing research to build an atomic bomb, asked Einstein to sign a letter to Franklin Roosevelt alerting him of the German effort and urging the president to initiate a similar project in the United States. Faced with an ethical dilemma, Einstein acted against his pacifist principles and signed the letter. He later confided that this was the one great mistake in his life, but there was some justification—that the Germans would make the bomb first.

In America Einstein continued his work on the unified theory and also became active in civil rights movements, fighting racism with passion and commitment and campaigning for the civil rights of African Americans. He also joined Jewish groups and supported Jewish causes. Near the end of his life, he said, "My relationship to the Jewish people has become my strongest human bond ever since I became fully aware of our precarious situation among the nations of the world." When Einstein died in 1955 at the age of seventy-six, American nuclear physicist Robert Oppenheimer described Einstein as being wholly without sophistication and wholly without worldliness, with a wonderful purity, at once childlike and profoundly stubborn.

Any attempt to describe and characterize Einstein's legacy would be necessarily confined in a phrase, a paragraph, or even a whole book filled with the same superlatives that have been used for others. Instead, we will focus on his work and make an effort to understand the meaning and significance of his scientific contribution.

The best place to start is Einstein's 1905 explanation of the photo-electric effect, which earned him the Nobel prize in 1921. The photo-electric effect is the observation that many metals emit electrons when light shines upon them. It was first observed by Heinrich Hertz in 1887. This photoelectric effect is not to be taken lightly. It means that you can start an electric current in a circuit simply by shining light on a metal plate. This is the basis for solar power today.

This phenomenon is actually rather simple. We all know that light warms up the surfaces of objects when it shines on them for some time. That means that the object receives energy from the beam of light. It makes sense that, at some point, the energy supplied by the light will cause an electron in the object to increase its kinetic energy and escape. This escape is much more likely to occur in a metal because, as we know, metals have a plentiful supply of free electrons in their structure. These are electrons that are not strongly bound to specific atoms and are free to move about inside the metal.

An explanation of photoelectric emission based on classical physics is very much like the commonsense explanation just described. Energy is transferred from light to the electrons in the metal. Low light intensities would not provide sufficient energy to increase the kinetic energy of electrons and cause them to escape the metal. As light intensity exceeds a certain threshold level, the energy carried by the light wave becomes greater, and some electrons will begin to escape. Even more electrons will escape when the intensity of light is increased further.

Amazingly, this is not exactly what the investigations showed. Experiments performed in 1902 by Philipp Lenard, a German physicist who had worked as assistant to Hertz, produced results that could not be explained by classical theory. For example, the number of electrons that escaped increased with light intensity, but their kinetic energies did not. The kinetic energies increased only with increasing light frequency. Blue light caused the emitted electrons to move faster than red light did. The dependence on frequency did not make any sense in classical theory.

In his groundbreaking 1905 paper, Einstein developed a theory that explained this unexpected result. Einstein solved the paradox using Planck's theory that light energy does not flow continuously but comes in little lumps, now called photons. He showed mathematically that the energy of emitted electrons increases linearly with the frequency of incident light and is independent of the intensity of light. Einstein's interpretation elevated the scientific value of Planck's theory, making it a cornerstone of modern physics. We do not know if Planck's long friendship with Einstein was in any way the result of Planck's gratitude.

Einstein's 1905 paper on the photoelectric effect is a masterpiece of mathematical logic and simplicity. Readers who are mathematically inclined will gain tremendous value from reading the original paper.[35] We will, however, resist the temptation to fill this page with mathematical symbols and equations and will opt instead for an understanding of the physical process.

We know that in a metal, there are electrons that are not strongly attached to specific atoms and are free to move around. These electrons feel attractive forces from atoms, but these forces cancel out as they come from all directions around the electron. If the electron is near the surface of the metal, it may just receive sufficient energy from the striking photons to acquire the kinetic energy required for an escape. When a photon strikes an electron, the photon disappears. It never had any mass, after all; it was all energy. The photon's energy is now absorbed in the electron as kinetic energy, and the electron moves faster. The energy received from the photon is proportional to the photon's frequency, as we know from Planck's law. If the frequency (and hence the energy) of the incident photons is high enough, as in ultraviolet light, more and more electrons will gain the necessary kinetic energy for an escape, including some of those electrons that

---

35 Einstein's original paper is titled *Concerning an Heuristic Point of View Toward the Emission and Transformation of Light* and can be downloaded from several websites.

are on the outer orbits of specific atoms. Increasing the intensity of incident light only increases the number of photons that collide with electrons, but the photons will not have sufficient energy to impart to the electron if the light has low frequency. The electron will gain some kinetic energy from the photon, equal to the photon's energy $E = hf$, where h is Planck's constant and f is the frequency of incident light. If this energy is not sufficient, the electron will move faster than before but will stay confined in the metal. Only increasing the frequency (color) of incident light will increase the electron's ability to escape beyond the threshold point and hence start the emission. This was actually confirmed in later experiments. It was also confirmed that light of low intensity but high frequency was able to start the emission.

We can now summarize the photon theory of light as follows: Light consists of small indivisible chunks of energy called photons, which move in a wavelike fashion at the speed of light. When a photon collides with an electron, the photon disappears. If the electron is a free electron, its kinetic energy increases. If it is a bound electron in an atom, the electron jumps to a higher orbit, and its potential energy increases. When the electron falls back to a lower orbit, it releases a photon. Each photon has energy equal to Planck's constant multiplied by the frequency of its wavelike motion.

A good way to visualize the photon's wavelike motion through space is to think of a snake trying to move quickly. If you drive on rural roads and have seen a snake trying to cross the road, you may have noticed that the snake transports its body in a wavelike fashion and moves forward along a path at a certain speed. The photon moves in a similar way but moves at the speed of light. Photons of blue light move in waves of higher frequency than photons of, say, red light, but both types move at the same speed—the speed of light. As a result of its higher frequency, blue light transports more energy than red light.

The photon theory of light, also called the quantum theory of light, does not replace classical electromagnetic theory. It adds to it in a way that makes possible the explanation of experimental results. This is again a demonstration of how preconceived ideas must change when empirical evidence proves them wrong or inadequate.

The annus mirabilis papers of 1905, Einstein's extraordinary year, included a paper titled "Does the Inertia of a Body Depend upon Its Energy Content?" This is the paper wherein Einstein developed his theory of equivalence of mass and energy, $E = mc^2$, the most famous equation in history. The paper is only two pages long and has no more than eight equations. The mathematical derivation is remarkably simple and uses no higher mathematics than algebra. Einstein concludes the following:

> If a body gives off the energy L in the form of radiation, its mass diminishes by L/c². The fact that the energy withdrawn from the body becomes energy of radiation evidently makes no difference, so that we are led to the more general conclusion that: The mass of a body is a measure of its energy-content; if the energy changes by L, the mass changes in the same sense by L/9 × 10²⁰, the energy being measured in ergs, and the mass in grammes. It is not impossible that with bodies whose energy-content is variable to a high degree (for example, radium salts), the theory may be successfully put to the test. If the theory corresponds to the facts, radiation conveys inertia between the emitting and absorbing bodies.[36]

Einstein says that a body that has radiated energy E has lost mass equal to E/c^2. Note that we use the more common symbol E for energy instead of Einstein's L. Therefore the mass lost is $m = E/c^2$ or equivalently $E = mc^2$. We realize, of course, that c^2 is a very large number. It is equal to the square of the speed of light, which in meters per second is the number 9 followed by sixteen zeros. (Einstein comes up

---

36 Einstein's original paper is titled *Does the Inertia of a Body Depend upon Its Energy Content?* and can be downloaded from several websites.

with twenty zeros, as he uses an older system of units based on the centimeter). This means that a tiny piece of mass has an enormous energy content. The simplicity of Einstein's language and mathematics is evident in this paper and refutes the old adage that says it is easier to understand the work of a great thinker from the writings of others. With Einstein's work our understanding is greatly enriched by reading the original papers.

One of the 1905 papers deals with the special relativity theory and is titled *On the Electrodynamics of Moving Bodies*.[37] This is the paper that outlines Einstein's ideas about the relativity of space and time. As with all of Einstein's work, we will focus on the development of the theory and the understanding of the physical concepts that underlie the mathematical formalism. Mathematics cannot be an end; it is a means to an end. As we are not mathematicians, we will not rid ourselves of the anxiety of assigning physical meaning to mathematical symbolisms. We will need to extract the scientific truth from the dryness of mathematical symbolism and make it part of our understanding, intuition, and common sense.

We may remember from high school how we used the Cartesian coordinates to describe an object's position in three-dimensional space with the three coordinates x, y, and z. The Cartesian system is an example of a reference system, and it is as useful in relativity as it was in classical physics. If our object of interest is moving, we simply track the changes of its Cartesian coordinates within the fixed Cartesian system. Nothing prevents us, of course, from thinking of a moving Cartesian system with a moving body inside. This is the case when we have a body that moves inside a moving train. In this case we have a fixed Cartesian system C, which is the stationary earth, as well as a Cartesian system C', our train, which is moving relative to C.

The earth, of course, is not stationary, but sometimes we will need to assume that our basic reference frame is stationary in order

---

37 Einstein's original paper is titled *On the Electrodynamics of Moving Bodies* and can be downloaded from several websites.

to explore certain phenomena. In relativity we will have to modify the Cartesian system somewhat by adding the coordinate "time." As it is not so easy to visualize a four-dimensional coordinate system (the three spatials x, y, and z, plus time) we will think of our familiar three-dimensional Cartesian system that moves in space as time passes or, better yet, moves in space-time. Einstein brings into relativity the Minkowski space, which is a four-dimensional space-time coordinate system proposed by German mathematician Hermann Minkowski. In much of our discussion, we will use the word "frame" instead of "reference frame" or "coordinate system."

Everything in the universe is in motion, and objects acquire their time dimension through their motion. Suppose for a moment that the universe consists of bodies that are fixed in space, and nothing is in motion. There is a distribution of matter throughout our imaginary universe, and matter consists of objects of different densities, such as rocks, planets, meteorites, dust, and air, but nothing ever moves. We can describe this universe completely by writing the space coordinates of each molecule of matter. Time does not exist, and it is superfluous in our description of the universe.

Let us now suppose that, for some reason, a molecule moves to a new position in space. It does not matter what the reason is. It could be a beam of light coming from somewhere that provides the necessary energy for the molecule's motion. Now that we have motion, we need the concept of time for a full description of our universe. The dimension of time is born with motion, which is nothing but a change of position in space. We still cannot measure time, because we do not have a periodic event that will help us define a unit of measuring time. But time does exist, because we now have several different conditions of the universe defined by the path of the molecule from its first to its final position. If we had a means of measuring time, we could even define velocity as the change in the molecule's position divided by the time taken to get there. So velocity is change in position per unit of time, and time is also defined only when there is a change in position. It seems that there is a great deal of similarity between time and velocity, and we will see a relationship between the two in Einstein's idea of time dilation.

Two of the most important results of special relativity are length contraction and time dilation. Length contraction means that when an object is moving at a constant speed with respect to an observer, its length in the direction of motion, as seen by the observer, is shortened by an amount proportional to the Lorentz factor, which is $1/\sqrt{1 - v^2/c^2}$ where v is the speed of the moving object and c is the speed of light. Time dilation means that when an observer is moving at a constant speed with respect to another observer, the clock of the moving observer appears to the stationary observer to tick more slowly than the clock of the stationary observer. The time difference registered by the stationary observer is proportional to the Lorentz factor.

The Lorentz factor is a relativistic relation that appears in much of Einstein's work and precedes Einstein's 1905 papers by about ten years. In fact, special relativity was initially called the Lorentz-Einstein theory for its dependence on Lorentz fundamentals. The Lorentz factor was developed in the 1890s by Hendrik Lorentz, a Dutch physicist who had developed a mathematical transformation to explain how the speed of light was independent of the reference frame and to understand the symmetries of Maxwell's equations.

We need to understand why time slows down at fast speeds, as this is an important outcome in special relativity. Let us think about how we measure time. We do so by observing and registering repeating events. For example, a pendulum swinging across back and forth is endlessly repeating the same swing and records one second of time per swing. If the clock is in the moving train together with an observer, the observer will see the pendulum swinging back and forth exactly the same as when the train was stationary. But a stationary observer standing on the station platform as the train passes by will see the train's clock pendulum traversing a greater distance from one end of the swing to the other. This greater distance is because the pendulum must travel across to complete its swing and at the same time must travel forward in the direction of the train's motion. The pendulum will still register one second of time, but the distance traversed will be greater. One second is therefore slower to the outside observer than one second when the train was stationary. The slowing of time is not perceptible at earthly speeds, but it becomes significant at very high speeds that are of the same order of magnitude as the speed of light. This slowing of time is called time dilation.

If we use a clock that works with light pulses instead of a mechanical pendulum clock, the result will be the same. This type of clock has a light source and light detector at its base and a mirror at a height h from its base. We assume that the clock is moving at a high speed along with an observer on its platform. The light source sends a light pulse that is reflected by the mirror and comes back to the light detector. The moving observer sees that the total distance traveled by the light pulse is $2h$. The time it takes for the round trip is $2h/c$, which is the distance traveled divided by the speed of light. The clock registers one second of time.

Let us now think that there is a stationary observer who stands by as the clock moves past. The stationary observer sees that the source sends a light pulse, but by the time it hits the mirror, the position of the mirror has changed somewhat in the direction of motion. The pulse is reflected back, and by the time it hits the detector, the position of the detector has changed some more, again in the direction of motion. The stationary observer sees that the pulse travels a longer distance than before, and therefore he or she registers a longer unit of time in his or her own apparatus than the moving clock does.

It is important to realize that time dilation is not an optical illusion and is not caused by the subjectivity of the human observer. It is caused by the relative speed of the two reference frames. The same time dilation would be recorded if, instead of a human observer, we had a clock or a photographic plate.

A logical consequence of time dilation is the loss of simultaneity or, more appropriately, "relativity of simultaneity." This is an important idea in special relativity. It means that events that are simultaneous to the observer on the station platform are not simultaneous to the observer on the moving train. The greater the speed of the train, the more pronounced the loss of simultaneity will be. Therefore, every reference frame has its own particular time. The statement of the time of an event has no meaning unless we are told the reference frame that the statement refers to.

Similarly, the idea of length contraction means that the length of a moving object, such as a rod, is subject to the Lorentz factor. The physical size and structural compactness of the rod certainly does not change, but the moving rod does appear shorter to the stationary observer, and the difference in length is proportional to the Lorentz factor. Length contraction is difficult to prove by direct experimental confirmation because objects of measurable length cannot be accelerated to relativistic speeds with current technology. The only objects traveling with near-relativistic speeds are atomic particles but their physical magnitudes are too small to allow a measurement of contraction.

Time dilation and length contraction are not optical illusions and are not a result of human subjectivity. We might call them measurement illusions or, better yet, relativistic effects. They must be taken into account when we make observations of objects moving at relativistic speeds. Lengths and times no longer have the absolute character attributed to them in Euclidean geometry and classical physics, respectively. Relativity tells us that the relative speed between object and observer must be accounted for because it changes the measurement of lengths and times. Time can no longer be regarded as independent of position and motion, and this is what makes it necessary for us to think in terms of space-time. A point in the Cartesian frame represents the position coordinates x, y, z of an object in space. A point in the four-dimensional space-time frame represents the existence of a physical "event" in space-time. Our focus in relativity has shifted from objects to events.

Has relativity replaced Newtonian mechanics? The answer must be positive for the microcosm, where we can have relativistic speeds at the atomic level. But the answer must be negative for the macrocosm, where the speeds of planetary objects are a tiny fraction of the speed of light.

Let us think of cosmological speeds for a while. As passengers on our planet earth, we travel along the earth's revolution about its axis at the speed of over one thousand miles per hour, if we are in the tropics. That is about twice the speed of a Boeing 747. If we are in New York, our speed is more like eight hundred miles per hour because our orbits become smaller as we move away from the equator. Those are good speeds—but nothing close to our orbital speed. On our orbit around the sun, we travel at sixty-seven thousand miles per hour. That's a better speed, but it pales in comparison with the speed of light. Our sun orbits around the center of our galaxy at forty-three thousand miles per hour, dragging all of us along at that phenomenal speed. At 107,000 miles per hour on its orbit around the sun, Mercury is the fastest planet. Thus, it turns out that Mercury is the fastest object in our solar system—quite appropriate for a planet named after the God of speed! But even Mercury's speed is only a tiny fraction of the speed of light—so tiny, in fact, that its Lorentz factor is 1.00000002, which means that any relativistic corrections to planetary speeds are meaningless, and Newtonian mechanics apply!

According to some scholars, Einstein's general relativity theory is the most beautiful physical theory ever invented.[38] The mathematical formalism is rather difficult as it uses tensor mathematics, curved space-times, and non-Euclidean geometries. This is advanced mathematical symbolism, and we will avoid it as much as possible. We are

---

38 Sean M. Carroll, *A No-Nonsense Introduction the General Relativity* (Chicago: University of Chicago, 2001), 2.

not Einsteins, but we should be able to translate the symbolism into our natural language in order to grasp the physical concepts.

Einstein had restricted his special relativity theory to "inertial frames," which refers to frames with bodies that are either at rest or are moving at constant speed. No forces were acting upon bodies, and therefore no acceleration was allowed. Gravity could not be accounted for, because gravitational fields create forces and accelerating bodies. Einstein embarked on a major effort in 1908 to include the effects of gravity in a more generalized theory. He published the theory in 1916 in a treatise titled *Foundations of the General Theory of Relativity*. A few months later, he published a small book titled *On the Special and the General Theory of Relativity Generally Comprehensible*. As its title suggests, the book was intended for a wider audience and was written with the minimum amount of mathematical language. We can plainly say that there is no better place than this book for a nonmathematician to learn about relativity. It is a must-read for anyone wanting good insight on the meaning of relativity; and thankfully, the book is in the public domain and downloadable at no cost from websites such as the Gutenberg Project website. Bertrand Russell's excellent little book titled *The ABC of Relativity* is another great resource that explains the theory in nonmathematical language and provides an excellent discussion of the philosophical aspects of relativity.

Despite all the recognition Einstein received for his special relativity theory, especially after Max Planck's endorsement and support, the new theory did not achieve immediate appreciation in the scientific community. That changed after Arthur Eddington's expedition to southern Africa and the successful empirical confirmation of the theory. The confirmation of the theory made worldwide headlines, and Einstein became an instant celebrity, the best-known person in world science.

The general theory of relativity is a theory of matter and gravity, and its mathematical formalism is complicated. That, of course, will not prevent us from trying to understand its physical meaning.

Everything in general relativity is interesting, but we will have to limit our review to a few outstandingly significant points. The first important idea is that light is deflected by gravity. This is not so difficult to understand, and we saw that it was confirmed in the Eddington expedition. But still, the idea is not intuitive. Why is light, which consists of photons without mass, deflected by gravity? What happened to Newton's law that gravity is a force of attraction between two masses?

A possible explanation might be that, due to the mass-energy equivalence $E = mc^2$, packets of energy behave like mass and display masslike characteristics in a gravitational field. But questions remain. Could something else other than gravity be happening in the vicinity of the solar surface that might be attracting light and be equally consistent with Eddington's result?

Remember, the solar surface is not a fun place to be, certainly not recommended for a winter vacation in the sun (pun intended!). It is a boiling environment in constant motion, full of helium and hydrogen atoms, free electrons colliding with everything on their aimless paths, naked atomic nuclei, loose protons and neutrons, magnetic fields, flares, and solar winds. Is it reasonable to expect that light photons passing through would leave unscathed, even in the absence of gravity?

So the basic question is this: why are massless light photons attracted by gravity? If we really want to answer this question effectively in relativity's favor, we can do it in a couple of different ways. First, photons may have zero mass, but they have energy, which in relativity is equivalent to mass. The photon therefore has inertial mass, which, by one of the postulates of relativity, is equivalent to gravitational mass. Second, relativistic gravity does not arise from the force of the gravitational field but rather from a geometrical distortion of space-time that occurs near large masses. These two answers may prove adequate if we just go through the math, but then another question arises: is this a pure theoretical construct, or does it also have a physical meaning?

The second important idea of general relativity is exactly the geometrical distortion of space-time suggested in the above paragraph. Space-time gets distorted and acquires curvature characteristics near large masses. This idea presents conceptual difficulties that arise, most likely, from our lack of empirical knowledge of space-time as a single frame. Our lack of direct experience of high velocities and curved, non-Euclidean geometries does not help, either. The need for non-Euclidean geometries arises from the curvature of space. Straight lines and perfect circles are Euclidean idealizations and do not really exist in nature.

Think for a moment of the meridian that goes from the equator through New York and ends at the North Pole. Then think of the meridian that goes through Athens, Greece. These two meridians, together with the equator segment between them, form a huge triangle, which has two right angles at its base, each being exactly ninety degrees. The triangle is completed with a one-hundred-degree angle at the top. The sum of the three angles is 280 degrees, far exceeding the 180 degrees required in Euclidean geometry that we all had to learn at school. This is the result of curved space.

A straight line drawn between two cities on a flat map is not the shortest distance between the two cities, and our commercial airlines know this quite well. The flight from Frankfurt to Seattle does not head directly to the west from Frankfurt. It travels northwest toward Greenland and passes over the southern part of Greenland before heading southwest toward Seattle. The shortest distance between two cities is a curved path; and similarly, the planets in the solar system follow the shortest paths, those that require the least amount of energy. The planetary motions we observe are the result of this economy. In their motion around the sun, the planets take a curved path because the huge solar mass bends space-time around it; and Euclidean geometry is not the most accurate mathematical tool in dealing with problems such as this.

Einstein's theory does not prove in any way that large masses cause a distortion of space-time. It just describes gravitational effects as distortions of space-time rather than as outcomes of physical gravitational forces. In general relativity, gravity is not a force, as Isaac Newton had proposed. Gravity is a consequence of the geometric distortion of space-time near large masses. Be assured that this is a difficult concept, and even physicists find it difficult to visualize what a distorted, curved space-time looks like. If we cannot visualize space-time, it is rather hard to visualize a distortion of curved space-time.

Common concerns regarding relativity are not necessarily a criticism or refutation of the theory. They arise from a disconnect between intuitiveness and plausibility. Certain aspects of relativity are far from being intuitive but are not necessarily implausible. The difficulty arises from the fact that we have learned to be intuitive only about ideas that are believable. Our intuition has been formed by cause-and-effect relationships that our empirical past has shown to be possible.

An extraordinary experiment took place in 2011 in the CERN nuclear research facility in Geneva. CERN is the European Organization for Nuclear Research that operates the largest particle physics laboratory in the world. The experiment found neutrinos that were moving faster than the speed of light. The results were published, and many scientists were shocked; others hailed the result as a refutation of relativity. The idea that the speed of light is the maximum speed in the universe is one of the cornerstones of relativity and modern physics. The difference in speed between the neutrinos and light was greater than that allowed by the statistical error of the experiment. Other labs around the world were unable to reproduce the same results. CERN replicated the experiment in 2012 and found that the neutrinos did not exceed the speed of light and that the first result was due to various measurement errors. What a sigh of relief for concerned prorelativity scientists!

The fact is that, one hundred years after Einstein's publications, the vast bulk of his ideas are standing up to the test of time. At the same time, hundreds of scientists around the world are still trying to find new experimental methods of verifying or refuting his theories.

Regardless of the failure or success of these tests, Einstein is the undisputed greatest scientific mind since Newton. Einstein's unique mind and vision have turned our attention to a very different and totally unexplored view of nature. He gave us radically original visions and concepts of natural phenomena and opened up scientific exploration in many new areas. The scientific research based on Einstein's ideas is as vigorous and enthusiastic today as it has ever been.

Quantum theory and relativity marked the dawn of a new age in man's understanding of the natural world. In our discussion of Einstein, we briefly mentioned his efforts to develop a general theory that would unify the forces of the universe and the laws of physics into one framework. Einstein felt very strongly that all of nature must be described by a single theory. He spent the latter part of his life at Princeton on this effort but was not successful in achieving a general theory. He was successful, however, in motivating thousands of other scientists into this *Road Not Taken* of scientific exploration.

A proposal was made by German mathematician Theodor Kaluza to expand Einstein's four-dimensional space-time into a five-dimensional space that would include space-time together with Maxwell's equations, uniting gravity and electromagnetism. Einstein was enthusiastic with the proposal and wrote to Kaluza that "the idea of achieving unification by means of a five-dimensional cylinder world would never have dawned on me. At first glance I like your idea enormously." Kaluza's idea was later abandoned, but Einstein never gave up his quest for a unified theory.

Einstein never quite accepted quantum theory, and his failure to develop a unified theory may be due to this fact. Near the end of his life, he found himself somewhat isolated from the scientific community and became absorbed in mathematical formalism, detached from the physical intuition that inspired his early discoveries. Sixty years after Einstein's death, unified theory is still the holy grail of modern physics.

The Anarchy of Science

Rand, Feyerabend, Kuhn

W E CAN COMPARE and confirm sensations received by different individual observers and thus form an objective view—that is, a view that is common to all observers and therefore void of observer idiosyncrasies. But this is as far as we can go with the notion of objectivity. We cannot observe the absolutely objective and true reality unless we step outside of our human perspective, which is impossible. In spite of our given skepticism of absolute objectivity, we cannot dismiss the new wave of objectivism in twentieth-century philosophy centered around the notion that reality exists outside of human consciousness—a view that seems quite logical to many of us but not as easily adopted by some philosophers.

The objectivist movement was founded by novelist, screenwriter, and philosopher Ayn Rand. She was born in 1905 in Saint Petersburg, Russia, to a middle-class family. Rand showed early promise in writing and mathematics. Her family was devastated in the 1917 Revolution, during which her father lost his pharmacy business. Rand enrolled at the University of Petrograd and studied history, politics, philosophy, and literature, but she was soon disillusioned with the suppression of free thought. At the age of twenty-one, she received permission for a brief visit to relatives in the United States, but her intent was to stay there permanently. Having studied Western history and culture, she was an admirer of America's individualism, vigor, and optimism.

Rand's first significant novel was not published until 1936. It was titled *We the Living* and was autobiographical of her early years, with vivid depictions of the brutality of the Soviet system. Intellectuals in post–Great Depression America were rather sympathetic to the Soviet experiment, and Rand's novel was not well received. Rand's next novel, *The Fountainhead*, was finally published in 1943 after being rejected by twelve publishers. The novel became a best seller and was even made into a Hollywood film starring Gary Cooper. The film was panned by critics but led to increased sales of Rand's book, making Rand famous. In the novel Rand uses a fictional plot and characters to lay out her groundbreaking philosophy of objectivism.

Another novel followed several years later, titled *Atlas Shrugged*, which was an immediate best seller and is, by most accounts, Rand's magnum opus and the most complete expression of her philosophical vision. Rand described the theme of the novel as the role of the mind in man's existence and, as a corollary, the demonstration of a new moral philosophy: the morality of rational self-interest. Rand's plot and characters are integrated into a comprehensive philosophy that includes metaphysics, epistemology, economics, and psychology. This was Rand's last work of fiction.

In the 1960s Rand began writing philosophical articles and essays in various nonacademic journals as her works were generally shunned in academia. Rand's philosophy had moral and political implications that were not in line with prevailing views of academic philosophers. Rand's own attitude was often disapproving of other philosophers, and they returned the compliment by dismissing her work.

Rand's concept of man is that of a heroic being with his own happiness as the moral purpose of his life, with productive achievement as his noblest activity and reason as his only guiding force. This concept reveals Rand's intellectual lineage: Aristotle-Aquinas-Nietzsche, a line of moral thought that we have not explored in this book, as our interest is centered on the philosophy of knowledge. Accordingly, we will focus here on that part of Rand's work that is relevant to our review of theories of knowledge.

That is not a simple task, as her epistemological propositions are often mixed with metaphysical, ethical, and political ideas. Rand acknowledged that Aristotle was the only philosopher who ever influenced her, although she had found inspiration in Aquinas and Nietzsche as well. She believed epistemology was a foundational branch of philosophy and considered the advocacy of reason to be the single most significant aspect of her philosophy.

Rand's best-known nonfiction book is the *Introduction to Objectivist Epistemology*, published in 1979 just three years before her death. Rand discusses the mental processes of conceptualization, the nature of definitions, distinguishing legitimate concepts from anti-concepts, the hierarchical nature of knowledge and what constitutes valid axiomatic knowledge.

Rand is definitely an empiricist as she believes that all knowledge is derived from perception. An idea can be validated only by tracing it to its source in perception. Reality consists of entities, which are the objects of our perception. Entities have a status that is independent of our conscience and a status that is defined by the entity's relation to our conscience. The question arises: does Rand believe that humans perceive both of these statuses?

We believe that the correct interpretation of Rand's theory is that humans form their own perception of reality while knowing that reality exists independently of human existence. This may be trivial and obvious to many of us, but it is not trivial to some philosophers who are still questioning the possibility of a real world that can exist without mankind. After 2,400 years, Plato still rules in many philosophical quarters!

Rand describes axiomatic concepts as the identification of a primary fact of reality, which cannot be analyzed or reduced to simpler parts. The three axiomatic concepts identified in her book are existence, identity, and consciousness. Existence is recognized as is, and an examination of the causes of existence is not meaningful. The concept of identity simply means that everything that exists has a specific, identifiable nature. The concept of consciousness recognizes the existence of consciousness in accordance with Descartes, that one cannot coherently deny the existence of one's own consciousness. Actually, this part of Rand's theory is a bit simpler than it appears: Rand recognizes existence, identity, and consciousness as axiomatic entities. They are a reality that we must accept without having to prove why and how they came to be.

Rand's ideas were influential during her lifetime and following her death. Her books have sold more than twenty million copies and continue to sell hundreds of thousands of copies each year. Excerpts from Rand's works are regularly reprinted in college textbooks, and several books have been published containing her early letters, journals, and other writings. Ayn Rand has not just won the public, she has made inroads into academia as well! That is quite a feat in the agnostic, skeptical world of modern academic philosophy.

In our exploration of philosophical theories of knowledge, our first goal is to find a good analysis and description of the process of scientific discovery. Our second goal is to find, if possible, a prescriptive system for scientific discovery—a kind of reasoning blueprint that will guide us through to a better understanding of natural phenomena. The question arises: has philosophy been helpful in the pursuit of these goals, or is it just an endless, meaningless discussion of unprovable and useless theories?

Paul Feyerabend, an Austrian philosopher born in 1924, would likely suggest that philosophical inquiry of the nature of knowledge and the scientific process has been useless. Initially trained as an opera singer and as physicist, Feyerabend studied philosophy under Karl Popper, became a critic of Popper's theories, and is now recognized as one of the twentieth century's most iconoclastic philosophers of science. In his book *Against Method*, Feyerabend rejected the existence of a scientific method and advanced his theory of epistemological anarchy, the idea that there are no useful rules governing the growth of knowledge and scientific discovery. The idea that science should operate according to universal methods is unrealistic and detrimental to science.

Feyerabend objected to any prescriptive scientific method on the grounds that any such method would limit scientific creativity and restrict scientific progress. Science would benefit from a dose of theoretical anarchism. Feyerabend showed that some historic events in science, such as the Copernican revolution, did not conform to scientific methods described and prescribed by philosophers. He claimed that the use of such methods would have actually prevented the discoveries.

Feyerabend is much closer to the scientific than the philosophical establishment. His criticism is aimed primarily at science philosophers who have been struggling for centuries to prescribe an effective scientific method, while brilliant scientists, although philosophically uninformed, continued to make advances with extraordinary new discoveries, proving that successful scientific research does not conform to any models prescribed by philosophers. The scientific process is complex, and philosophy cannot devise methods of differentiating science from pseudoscience or myth.

We will borrow the following excerpt from the *Stanford Encyclopedia of Philosophy* reporting an interesting incident during a Feyerabend lecture: "His listeners were enthralled, and he held his huge audiences until, too ill and too exhausted to continue, he simply began repeating himself. But not before he had brought the house down by writing 'Aristotle' in three-foot high letters on the blackboard and then writing 'Popper' in tiny, virtually illegible letters beneath it!"[39]

39 http://plato.stanford.edu/entries/feyerabend/#2.13.

Feyerabend's work shifted the focus of the philosophy of science away from the verificationism of the logical positivists and from Popper's falsifiability criteria.

This shift would take another turn toward a historical, evolutionary perspective with the work of Thomas Kuhn, an American physicist, historian, and philosopher. Kuhn was born in Cincinnati in 1922 and studied physics at Harvard, where he developed a strong interest in the philosophy of science. Kuhn is probably the most influential philosopher of science in the twentieth century. He taught history and philosophy of science at Harvard, Berkeley, Princeton, and MIT. In 1957 he published his first book, *The Copernican Revolution*, wherein he studied the development of the heliocentric theory of the solar system during the Renaissance. Five years later he published his landmark work, *The Structure of Scientific Revolutions*, which offered a brand-new perspective of the evolution of science.

Kuhn was the first author to articulate an alternative to the traditional view of scientific progress. In Kuhn's terms, "normal science" is the regular experimental work scientists conduct within a given paradigm, a given framework of theories, rules, and assumptions. Normal science is based on the assumption that the scientific community knows what the world is like. Normal science works within the existing scientific paradigm. Working within the current paradigm framework, scientists generally devote themselves to solving scientific puzzles. Their solutions reinforce and extend the scope of the paradigm without much interest in challenging it. When experimental results fail to conform to the existing paradigm, anomalies are created, which are often ignored until they accumulate, develop into a crisis, and lead to a scientific revolution, which establishes a new paradigm with new rules and theories. This last phase is called "revolutionary science," and the transition from the normal to the revolutionary phase creates a "paradigm shift," an expression that was not coined by Kuhn but was popularized by Kuhn and is in wide use today in many disciplines, including business. Its use has actually become a cliché.

One of the most important concepts in Kuhn's theory is the idea of incommensurability. That is a long, fancy word, but it has a simple meaning—that different paradigms have no common standards of comparison. The languages of the two theories lack sufficiently overlapping meanings, and their conceptual frameworks are so different that scientists are unable to use empirical evidence to compare one theory with the other. There are accounts of reality in the new paradigm that cannot be reconciled with certain aspects of the old paradigm. The idea of incommensurability was actually introduced in 1962 by both Kuhn and Feyerabend independently. It was probably an outcome of their discussions, as the two were close friends. Kuhn suggests that the proponents of each paradigm see the world in their own way because of their scientific training and prior experience. They use a different conceptual framework and have different ideas about scientific standards.

One of the problems in Kuhn's theory is that normal science and revolutionary science are considered distinct and apart from each other. One follows the other in succession. But science researchers know that normal and revolutionary work coexist and interact at all times, even within the same research project. They are not distinct from each other, as there are revolutionary elements in normal research and normal elements in revolutionary research. In fact, no one knows how to distinguish normal from revolutionary research. Scientific breakthroughs have occurred by accident during normal research. Most scientists believe there is a revolutionary element in their own research, and quite often the outcome proves them right.

Kuhn paints an overly conservative picture of scientists, who, in his view, are engaged in puzzle solving within the existing paradigm. But we know that successful scientists are able to think out of the box and inside the box at all times. Natural phenomena that are discovered and cannot be explained within the existing paradigm set the stage for the creation of a new theory, which cannot appear from nowhere but must arise from conflict within the existing paradigm. The history of science shows that the seeds of revolutionary ideas are always present, and it takes a brilliant mind to make them produce the fruits of great new knowledge. The outcome of a scientific effort determines whether the effort revolutionized science or not. After all, what is a scientific revolution? Kuhn does not define the term. Its meaning in Kuhn is derived by implication from considering the entire theory. We will propose that a scientific revolution is simply a discovery that opens multiple paths of new research. We will discuss a few examples of scientific revolutions to see where they conform to Kuhn's model and where they do not.

Newtonian physics and relativity are often cited by philosophers as being two competing theories. But Newtonian physics is not really in competition with relativity or contradictory to relativity. Relativity applies to objects moving at speeds close to the speed of light whereas Newtonian physics applies to objects moving at much lower speeds. In our discussion of relativity, we saw that relativistic time and length differ from their classical counterparts by the Lorentz factor, which is a correction factor that depends on the velocity of an object relative to the speed of light. The relative velocity of motion of two objects is the same concept in classical physics and relativity. The concepts are the same, and the mathematical language is the same. It is not conceivable that a trained physicist who specializes in relativity will have a problem communicating with another trained physicist who practices mostly classical physics. There are no such distinct categories of physicists or scientific communities.

It would be extremely difficult, if not impossible, for a physicist to prove anything in relativity work if the use of classical concepts were disallowed. Let us think for a moment how Einstein's brilliant revolutionary mind developed $E = mc^2$. The derivation is remarkably simple and uses concepts and formulas from Newtonian mechanics, such as momentum, kinetic energy, and the laws of conservation of energy and momentum. That is as classical as physics can be. It also uses Planck's and de Broglie's laws, which may be considered laws of the new paradigm but are still expressed in mathematical language that is fully understood in the classical framework. And most importantly, Einstein's derivation is based on the assumption that the laws of physics are the same in all reference frames! The speed of light is also the same in all frames! Einstein's derivation shows clearly that relativity's genetic origins are rooted in Newtonian physics. The common concepts and common mathematical language offer a standard for comparison, and we cannot claim that the two paradigms are incommensurate.

It is often said that the Newtonian theory of gravitation is not compatible with Einstein's general relativity. Newton believes that gravity is caused by a force, whereas Einstein's theory tells us that it is caused by a geometric distortion of space-time. Does relativity reject the existence of a force? There are physicists who believe so, but there is hardly any physicist who rejects the existence of nuclear forces, the forces that keep the atomic nucleus together. The concept of force, therefore, is not obsolete in modern physics. The force of gravity is something that we experience every day. In fact, this force can produce work, which is the most quintessential element of the concept of force. The apparent incompatibility between Newtonian gravity and relativity is resolved when we consider that the Newtonian gravitational force is an effect of gravity, which is a property of large masses caused by a geometric distortion of space-time. The preceding sentence is a clear statement of two cause-and-effect relations: the force is caused by the property of gravity, and the property of gravity is caused by a geometric distortion of space-time near large masses. We have two successive orders of causal relations and the idea of incompatibility of the two theories must be refuted.

Coming now to special relativity, if we consider the speed of an object as "context," then the differences between Newtonian physics and relativity are contextual, or frame-related. Structures, buildings, and machines were built successfully before anyone knew that $F = ma$ or $E = mc^2$. Satellites orbiting the earth were designed with the application of classical physics and without the knowledge of any relativistic principles. Those satellites were launched successfully and are still operating.

It is clear that Newtonian physics is not applicable to velocities that are of the same order of magnitude as the speed of light, and relativity rules in those instances. But we must remember that relativity, although widely embraced by the scientific community, has not been fully validated, at least not in all of its different parts. For a full validation we need technology that is not available right now. We need the ability to accelerate objects to velocities that are comparable to the speed of light. These objects cannot be tiny particles. They must have size and mass such that the minutest changes in mass are measurable. There is no technology currently available that can do this. Certain parts of relativity, therefore, are unverifiable and unfalsifiable with current experimental capabilities. In any event, the simple fact that velocity offers a standard of comparison between Newtonian and relativistic physics means that the two paradigms are commensurate.

The comparison of heliocentrism and geocentrism is another example that does not confirm Kuhn's theory. The Aristarchean heliocentric theory coexisted with Aristotle's geocentrism for several centuries, albeit not with the same popularity as Aristotle's system. The Copernican paradigm has many connecting threads both with Aristarchus and with Aristotle. The irony is that Einstein's theory of relativity upset both models. New evidence has also shown that the solar system's center of gravity is not the exact center of the sun. This means that either model is acceptable regardless of the fundamental differences between the theories. Astronomers use both the heliocentric and geocentric models for research, depending on which theory makes their calculations easier. They use reference frames with the origin in the center of mass of the earth, or the earth-moon system, the sun, or the sun-planets system. Astronomers will often mix, in the same study, heliocentric velocity and momentum with geocentric coordinates. Their selection of geocentric or heliocentric frames is merely a matter of convenience and, in the final analysis, it is only an approximation of the actual space-time. Their choice is made for ease of computation and does not have any great philosophical implications. If astronomers can work in both paradigms in the same study with such ease in their daily work, then we may say that the paradigms are commensurate.

Kuhn's work has earned both praise and skepticism. The traditional view of the evolution of science is that science progresses sometimes in an orderly fashion and other times in a haphazard fashion, constantly fusing old and new concepts. If there are accounts of reality in the old paradigm that cannot be reconciled with the ideas of the new paradigm, they will either be discarded or modified. If they are modified successfully, they will become part of the new paradigm. They will likely influence concepts of the new paradigm. Doesn't this fusion process remind us of the dialectical triad thesis-antithesis-synthesis that we saw in our discussion of Hegel? The Hegelian view of scientific progress is more intuitive than Kuhnian theory and is generally supported by scientists, whereas Kuhnian theory has been quite popular among philosophers. There is no doubt that Kuhn's influential work has fostered the development of a new area of study, the philosophy of science. Canadian philosopher Ian Hacking asserts that "one of Kuhn's marvelous legacies is science studies as we know it today."

Quest for the Holy Grail

Standard Model and Theory of Everything

N HIS INTRODUCTION to *The Feynman Lectures on Physics*, physicist Richard Feynman writes:

> I think I can safely say that nobody understands quantum mechanics. So do not take the lecture too seriously, feeling that you really have to understand in terms of some model what I am going to describe, but just relax and enjoy it. I am going to tell you what nature behaves like. If you will simply admit that maybe she does behave like this, you will find her a delightful, entrancing thing. Do not keep saying to yourself, if you can possible avoid it, 'But how can it be like that?' because you will get 'down the drain', into a blind alley from which nobody has escaped. Nobody knows how it can be like that.

Feynman did not receive the 1965 Nobel prize for being the great educator and popularizer of physics that he was, but his introductory passage is quite telling: physics is a description of nature; don't worry about the complexity; enjoy its fascinating beauty.

Difficult or not, we will strive to achieve a basic conceptual understanding of the new science while avoiding as much of the mathematical formalism as possible. Sometimes we may have to use uncommon sense to understand strange phenomena at the atomic level. Atoms are not like planets orbiting around the sun and do not obey Newton's or Kepler's laws. Their positions, velocities, and motion are different. We have already seen that the theories of Planck, de Broglie, and Einstein opened multiple paths of new research, which hit the scientific world like an avalanche. The interaction of light and mass became the new focus and an entire new area of physics developed, called quantum electrodynamics (QED). It is the quantum equivalent of classical electromagnetism. QED is the first theory that successfully combined the principles of quantum physics with special relativity, whereas quantum theory and general relativity are still incompatible, and we will see further down what efforts are made to achieve agreement between the two theories.

English physicist Paul Dirac is one of the founders of QED. Born in 1902 in Bristol, Dirac studied electrical engineering and mathematics at the University of Bristol. Unable to find work as an engineer, Dirac decided to do graduate work in physics at Cambridge, focusing on quantum theory and relativity. These fields of study were so new in the 1920s that there was only one professor at Cambridge who could supervise Dirac's graduate work in quantum physics. In 1926 Dirac developed the first complete mathematical formulation of quantum mechanics and then focused his energy on a relativistic formulation of quantum theory.

The rest of Dirac's story is truly amazing! In 1927 he developed a quantum theory of radiation, where he came up with a brand-new property called the electron spin, a measure of the electron's angular momentum, which eventually became one of the fundamental concepts in particle physics. This was a spectacular achievement, but Dirac knew from the very beginning that his theory had some serious problems. The mathematical equations yielded solutions that generally made sense. But there were certain credible conditions where the solutions did not make sense, as they required the electrons to have negative energy. Nobody knew what negative energy might be. You either have energy or you do not; you cannot have negative energy. Dirac was aware that a couple of years earlier, a Swiss physicist named Wolfgang Pauli had proposed the exclusion principle, which said that it is impossible for two electrons of the same atom to have the same values of the four quantum numbers.

The four quantum numbers are (1) the principal quantum number, which is the energy shell of the electron (that is, the electron's orbital level); (2) the azimuthal quantum number, which is the magnitude of the electron's orbital angular momentum; (3) the magnetic quantum number, which is the projection of the orbital angular momentum along a specified axis; and (4) the spin projection quantum number, which gives the projection of the spin angular momentum along a specified axis. These four numbers are the set of acceptable solutions to the Schroedinger wave equation for the hydrogen atom.

Pauli's principle was an unexpected stroke of luck for Dirac, who could now consider that the unoccupied quantum states were holes with a positive charge. He thought of these as "antielectrons." In other words, the forbidden positions were actually occupied by particles of the same mass as the electron but of the opposite electrical charge—a positive electrical charge. This is how the idea of antimatter was born. Now, if you think that this is too much speculative theory proposed for the sole purpose of making mathematical equations work, you will not be too far off the mark! But in another stroke of luck for Dirac, American physicist Carl Anderson accidentally discovered the antielectron in cosmic rays. It was a particle with the same tiny amount of mass as the electron but with a positive charge. This particle was named "positron." The scientific community was astonished, and Dirac's theory was the talk of the day. Dirac was awarded the Nobel prize in 1933.

This is all quite amazing, but the questions arise: Are we forcing nature to conform to a preconceived mathematical formalism? Are we forming perceptions of microcosmic reality that confirm predetermined and arbitrary notions? Is this a type of circular logic where we discover what we want to discover? Is physics on a course of becoming a self-fulfilling prophesy? Does nature have to conform to mathematical symmetry? Is mathematical beauty a necessary, sufficient condition for physical truth?

The next events in Dirac's life point to a negative response to these questions but that does not mean that the questions must be discarded. They are very legitimate questions, and we must be aware of them at all times. In a 1931 paper titled *Quantised Singularities in the Electromagnetic Field*, Dirac predicted the existence of another particle, the magnetic monopole, but no such particle has been found. More than eighty years have passed, and magnetic monopoles have not been found in nature, and it has not been possible to create them artificially.

Dirac's greatest contribution is that his QED equations describe for the first time the motion and spin of the electron in complete consistency with both quantum theory and special relativity. QED rests on the idea that charged particles, such as electrons and positrons, interact by absorbing and emitting photons, the particles of light that carry electromagnetic radiation. These are virtual photons that do not exist outside of the interaction and only serve as carriers of momentum. These interactions are described intuitively by Feynman diagrams.

Everything that we have discussed in modern physics seems to point to a discrete rather than a continuous reality. Energy comes in tiny discrete packets called quanta. Electrons jump from one orbit to another, and they simply do not exist in the space between orbits.

Their path from one orbit to another is undefined. It is almost as if they are destroyed on one orbit and instantly recreated on another. They have discrete energy levels, discrete spin, and angular momentum. There is no continuity and values between these quantized levels are forbidden, as if they do not exist. Pauli's exclusion principle is all about quantized states. Maybe some theory will emerge that shows space-time itself to be quantized. No wonder it is called quantum physics!

The quantum vision is not just a reality of discrete quantities; it is also a stochastic reality, a reality of probabilities and randomness. This randomness is actually what kept Einstein at a distance from quantum theory and made him declare that God doesn't play dice! Einstein was the supreme determinist. He said that "everything is determined by forces over which we have no control. It is determined for the insect as well as for the star. Human beings, vegetables, or cosmic dust, we all dance to a mysterious tune, intoned in the distance by an invisible piper."

In our discussion of Einstein, we saw that he spent the last twenty years of his life trying to develop a unified field theory, a general theory that would unify the forces of the universe and the laws of physics into one framework. He was not successful, and his lack of success may be partly due to his skepticism toward uncertainty and quantum mechanics. However, Einstein's commitment to the unified theory motivated other physicists and opened up a worldwide multilevel effort of new research.

The quest for a general theory has engaged thousands of scientists around the world on multiple, wide-ranging paths of new scientific inquiry. There is an important outcome that has filled many physicists with a renewed enthusiasm for the final solution—and others with intense skepticism. It is called the standard model. Like the Schroedinger equation and many other good things in modern physics, the standard model has been formalized in different versions. They all have the same variables and the same relations between variables, and they all have the same underlying principle. Now, this is some serious physics. The standard model rules in physics today, and we must know at least what it is. There is a side benefit in this discussion, as we will get acquainted with some interesting little particles that we have not yet talked about.

There are two types of particles in the standard model: particles that make up matter, called fermions, and particles that transmit forces, called bosons. In the class of fermions, we have the leptons and the quarks. There are six leptons: electron, muon, and tau, and their three counterparts, electron neutrino, muon neutrino, and tau neutrino. In the class of quarks we have six types: up, down, charm, strange, top, and bottom. The bottom quark is also known as the beauty quark, for reasons unknown. Remember, physicists are the type of folks who can find poetic beauty in the Pythagorean theorem.

Bosons are hypothetical particles the standard model requires to explain the transmission of four types of forces: the strong nuclear force, the weak nuclear force, the electromagnetic force, and the gravitational force. Of these forces, gravity is the weakest but has an infinite range, albeit with a strength that declines with the square of the distance, as we know from high-school physics. The electromagnetic force also has an infinite range but is much stronger than gravity. We already know what the boson of this force is: our familiar photon! The wave particle that light is made of, the energy quantum that carries electromagnetic energy. The weak and strong nuclear forces have a very small range and are required to keep the nucleus together as a compact unit. The weak force is carried by the W and Z bosons, and the strong force is carried by the gluon. As its name suggests, it keeps protons and neutrons glued together in a compact nucleus.

So we have described the bosons that carry three of the four fundamental forces in the universe, but what about gravity? standard model physicists assume there is another particle—the graviton—that carries gravitational force. This particle has never been detected in any experiments, and gravity is still not part of the standard model.

Newtonian gravity is not included, and neither is Einstein's distortion of space-time around large masses. This is actually the model's most serious deficiency right now.

It is quite amazing, if you think about it. The one force that we are all familiar with and experience in our daily life cannot be explained by the new physics! But there is a good reason for this. The standard model is tuned to microscopic phenomena, where gravity is so weak as to be unimportant. Gravity dominates in the macrocosm but is a negligible force in the microcosm. Does this mean that the standard model cannot become a theory of everything?

You might guess that the standard model will prove much better at explaining microcosmic phenomena than macrocosmic phenomena. This is not so strange if we think that the model is deeply rooted in quantum mechanics. But what is the difference between the macrocosm and the microcosm? Nothing, really, except that humans stand between the two. Everything that is visible to us belongs in the macrocosm, and everything else is in the microcosm. It is a difference of scale and a difference of human perspective. But the universe is a continuum of space-time that does not care about human perspective. So the distinction between macrocosm and microcosm is not a fundamental distinction of the universe but rather a distinction of human perspective. It may well be that the two can be related with one another through a relativistic reference frame—an observer-dependent frame.

The standard model is an impressive piece of physics. But the cockpit of a Boeing 747 is also impressive if you know what every switch does—and even more impressive if you do not know what anything does. While searching for a unified theory, a search that led to the standard model, physicists found a whole bunch of new particles. In other words, the search for simplicity led to more complexity. The diversity that we observe in nature when we watch those stunning *National Geographic* films seems to have an underlying diversity in the microcosm of elementary particles. Can all these new particles and their interactions fit into the straitjacket of a single equation or a single law of physics?

We know today that stars can be created and destroyed. Elementary particles are also created and destroyed. Tiny mass particles are annihilated when converted into energy packets. In our ever-changing, dynamic universe, is it possible that new laws of physics are randomly created and destroyed?

The mathematical formalism of the standard model belongs in an advanced textbook of modern physics, and we will resist the temptation to duplicate it here. The mathematics looks intimidating, mostly because of all the Greek letters and the multitude of terms. It is actually much simpler than it looks. It is basically a polynomial of linear and nonlinear terms. The problem is not with the formalism but with the conceptualization. The great details of interactions described in the model convey a sense of arbitrariness. We cannot explain something, so we invent a new term (a new particle), stick it somewhere in the equation, and we are done! Well, that is an exaggeration; it is not exactly like that. Physicists may be inclined to speculate more than the rest of us, but speculation is an important element in scientific discovery.

There are skeptics and physicists who are trying to shatter the standard model and build a brand-new theory. As it stands now, the standard model is a work in progress. Some of its terms, such as the terms relating to the Higgs boson, have yet to be verified. The standard model is not scientific fact; it is still a theory that needs to become verifiable and falsifiable. It has predicted the existence of unknown particles that have since been discovered, but there are serious gaps that still need to be filled.

The standard model is unlikely to become a theory of everything, and it may no longer be the holy grail of physics, but it has achieved rock-star popularity nevertheless. There are T-shirts and coffee mugs with the equation printed on them. But even without the standard model, physics continues to produce new incredible discoveries that reveal unknown aspects of our world while making our lives much more interesting through exciting practical applications. Lasers, transistors, superconductivity, high-definition TV, wireless communications, and laptop computers have changed our lives forever.

Billions and Billions of Stars

White Dwarfs, Black Holes, Red Giants, and Big Bangs

W E OFTEN SAY that life is too short, but there are species that have a life-span of less than one day. Other species, like houseflies, are much more fortunate, as they can live up to four weeks. As much as we think that their sole purpose in life is to irritate humans, houseflies do not really care about that. All they want is to eat and breed. Our familiar ants can live to three months, and humans live to one hundred years, more or less. But humans are not even close to the top of the life-span pyramid. Whales and Galapagos tortoises can live up to two hundred years. No one really knows if they spend all that lifetime on artistic, scientific, or other intellectual pursuits or whether they find the time to care about morality and religious beliefs. They probably just want to eat and breed, and they can keep doing this for two hundred years.

An ant's viewing range is no more than one meter. So an ant's perspective reality is only a tiny portion of our human-perspective reality. There is no way that an ant will ever know Paris or the Parthenon or

Niagara Falls. The ant's universe is limited to its colony, and it is only a tiny fraction of the universe known to humans. But isn't it likely that the universe that we humans can ever know is also a tiny portion of the entire universe? Isn't it likely that our human-perspective reality is limited, just like the ant's perspective reality is limited, by our sensory range and by the limits of our intellectual capabilities? No one really knows the answer, but we might speculate that, yes, it is more than likely that we can only know a tiny portion of the universe.

Scientists say that the Big Bang occurred 13.82 billion years ago. They do not even round the number to 14 billion, it is absolutely and precisely 13.82 billion years. So humans, with their one-hundred-year life-span and less than one hundred generations of accumulated scientific knowledge are able to discover evidence from 14 billion years ago. It really sounds strange, and yet this is the object of much scientific research in physics and astronomy today. But the travel back to the time of the Big Bang can be fascinating, even if we think that some of it is fiction. We will not see any witches or dragons along the way, but we will get acquainted with some incredible stories and sights of black holes, white dwarfs, red giants, and quasars. These are the witches and dragons of the universe.

Back in 1919 a thirty-year-old man named Edwin Hubble arrived at Mount Wilson in California to make some astronomical observations of the Milky Way, which at that time was considered to be the entire universe. Hubble was an unknown scientist at that time, with just a few years as a professional astronomer. He had studied mathematics and astronomy in Chicago and law at Oxford, but his studies had been interrupted by his voluntary enlistment in the US Army during World War I. Hubble's arrival at Mount Wilson coincided with the completion of the Hooker telescope, the world's largest at that time.

Hubble's observations in 1923 showed that certain nebulae, those cloudy patches that we see up in the night sky, were too distant from our galaxy and were, in fact, galaxies of their own. Andromeda, the princess of all nebulae, was one of those galaxies. While the scientific community was still in shock and disbelief that our treasured Milky Way is not the entire universe, the New York Times published an article titled *Finds Spiral Nebulae Are Stellar Systems. Doctor Hubble Confirms View That They Are "Island Universes" Similar to Our Own.*

Hubble's seminal moment was still ahead of him. From the spectra of the emitted light, he began to measure the velocities of all the known nebulae. After successive measurements of the same nebulae, he made a startling discovery: the spectra of the nebulae displayed a redshift over time. A redshift happens when light is shifted to the red end of its spectrum, meaning that its frequency is reduced. Does this remind us of the Doppler effect? It sure does! The nebulae were moving away from us. This was a tremendous breakthrough in astronomy as it overturned the conventional view that the universe is static. This is how the theory of the expanding universe was established.

It is interesting to note that when Einstein developed his general relativity, he found that his theory required the universe to be either expanding or contracting. He was puzzled by this result and he felt compelled to add a fudge factor in the equations in order to get rid of the problem, as the prevailing view at the time was that the universe is static. This fudge factor is known as the cosmological constant, and it is usually denoted by the Greek capital lambda (Λ). It is the energy density of the vacuum of space and it is meant to counteract the force of gravity. When Einstein learned of Hubble's discovery, he realized that the expansion predicted by his own theory was real, and the cosmological constant was redundant after all. Later in his life, Einstein said that changing the equations was the biggest blunder of his life. British astrophysicist Stephen Hawking wrote in his book *A Brief History of Time* that Hubble's discovery that the universe is expanding was one of the great intellectual revolutions of the twentieth century.

Hubble remained active at Mount Wilson for the rest of his life. He was never awarded the Nobel prize in physics, as astronomers at that time were not eligible for the prize. The space telescope launched by NASA in 1990 on a low earth orbit was named after Edwin Hubble. It is known as the Hubble Telescope! After an initial problem with one of its mirrors, the telescope was repaired in 1993 and started producing spectacular images of distant stars, nebulae, and galaxies.

General relativity predicted the existence of space-time deformities around dense masses behaving like black holes, from which no light or any other radiation can escape. The idea goes back to 1783 when an English clergyman and natural philosopher named John Michell wrote a letter to Henry Cavendish, a scientist and prominent member of the Royal Society:

> If the semi-diameter of a sphere of the same density as the Sun were to exceed that of the Sun in the proportion of 500 to 1, a body falling from an infinite height towards it would have acquired at its surface greater velocity than that of light, and consequently supposing light to be attracted by the same force in proportion to its inertial mass, with other bodies, all light emitted from such a body would be made to return towards it by its own proper gravity.[40]

In one amazing paragraph written 230 years ago, Michell described black holes; and by all accounts, he seems to have been the first person to do so. He called them dark stars. The theory was ignored for a long time, as it was not understood how light could be affected by gravity. Michell has been called "one of the greatest unsung scientists of all time."[41]

In addition to being the first person to propose the existence of black holes, Michell was the first to suggest that earthquakes travel in waves, the first to explain how to manufacture artificial magnets, and the first to apply statistics to the study of the cosmos, recognizing

---

40 John Michell, Letter to Henry Cavendish, *Royal Society Journal*, 1784, URL = http://www.relativitybook.com/resources/Michell_1783.html

41 http://www.exnet.com/1996/02/20/science/science.html.

that double stars were a product of mutual gravitation. Michell also invented an apparatus to measure the mass of the earth. He has been called both the father of seismology and the father of magnetometry.

The first thing that comes to mind about a black hole is that because it does not emit any light, it can never be observed. Well, if there is an unfalsifiable theory, this must be it! We have managed to come up with something that can never be discovered. We should perhaps recall Wittgenstein's words: "What can be said at all can be said clearly, and what we cannot talk about we must pass over in silence."[42]

Our philosophical skepticism aside, it was reported in May 2012 that the first visual proof of existence of black holes had been achieved. A team from Johns Hopkins University made observations through the Hawaiian telescope Pan-STARRS 1 and recorded images of a supermassive black hole 2.7 million light-years away that was in the process of swallowing a red giant![43] We do not know whether or not they showed this on TV as live entertainment!

So, the red giant was trapped like a fly in a spider's web and was then swallowed in one big gulp! The colorful description is really quite funny, but can it also be true? First of all, what is a red giant? It turns out that a red giant is a star in the latter part of its life, when most of its hydrogen, the most common element in the universe, has been converted to helium, in that familiar nuclear fusion that makes our own sun give us all that sunlight and warmth. Some of the most familiar stars in the night sky, like Aldebaran and Arcturus, are red giants. Another one is Betelgeuse, one of the four corners of the Orion constellation, a very familiar and visible star pattern on our sky. Astronomers say that our sun will also start to run out of hydrogen in about 5.4 billion years and will turn into a red giant.

There is no sure way of knowing how often these black hole feastings occur. Some astronomers say they are not frequent at all—just

42 Ludwig Wittgenstein, *Tractatus Logico-Philosophicus*, Project Gutenberg website, URL = http://www.gutenberg.org/files/5740/5740-pdf.pdf

43 http://www.scientificamerican.com/article/black-hole-swallows-star/.

one per galaxy every ten thousand years or so. But on the cosmic time scale, isn't this like putting away three meals a day? This stuff is more exciting and entertaining than the best science fiction ever written, but it is also serious business, and we need to get serious.

Suvi Gezari was the Johns Hopkins team leader who made the observations. She and her colleagues used a number of different telescopes to watch the black hole devour the red giant that had dared to come so close. This all happened in a galaxy two billion light-years away. In other words, it happened two billion years ago, and we just saw it in 2012. Things that are happening in the universe now will be seen a few billion years later, if there are still any intelligent beings watching the sky from somewhere.

Gezari's team coordinated its observations with simultaneous observations from NASA's Galaxy Evolution Explorer, which is an orbiting space telescope that makes observations at ultraviolet wavelengths to measure the history of star formation in the universe 80 percent of the way back to the Big Bang. Gezari and her team were able to analyze the constitution of debris matter from the consumed star. The material was found to be mostly helium with no hydrogen, a finding consistent with the red giant formation theory. This is actually how they determined that the devoured star was a red giant.

In any event, it may be premature to claim that the May 2012 astronomical report confirms the existence of black holes. It describes a celestial observation in a way that could be explained by the black hole hypothesis. But there could be other hypotheses, yet unproposed, that could explain the event just as well. The theory of black holes is very exciting stuff, but it is still an unconfirmed theory in a dark area of modern astronomy.

The space between stars is not empty; it is filled with clouds of gas and dust. These clouds are mainly hydrogen, which is the most abundant element in nature. Clouds can also contain helium and small amounts of other elements, including lithium, which is gradually destroyed during the nuclear fusion. That is why the presence of lithium in the light spectrum is a good indicator of a star's age.

If the cloud collides with other clouds or celestial objects, it may contract and develop a sufficient gravitational pull as it forms a mass of hydrogen at its center. The mass gets larger, the force of gravity increases, and, as the mass gets even larger, the contraction becomes self-sustaining. This is not a quick process and may take thousands of years. As more hydrogen collapses toward the center, the temperature and pressure at the core increase; and at a certain temperature, nuclear fusions begin to occur, creating helium and various other elements from the compression and fusion of hydrogen atoms. The process releases enormous energy. At some point the outward pressure from the fusion reactions will counterbalance the pull of gravity and prevent any further contraction. The star is now stable. It has reached maturity and is entering the main sequence of its life cycle. This is when the star is born!

Some cloud formations cannot achieve the large mass and high temperatures required for the fusion process and will not become stars. These are known as brown dwarfs. They will shine dimly; they may be attracted by a star and become planets, but they will eventually die over hundreds of millions of years. Our sun is a yellow dwarf and is thought to be in the middle of its life-span. When the hydrogen fuel that powers its nuclear reactions begins to run out, our sun will expand, cool down, and become a red giant.

We do not need to worry just yet, as this will not be happening for another five billion years or so. Small stars like our sun will eventually die a relatively peaceful death, passing through a nebula phase and then a white dwarf phase. A white dwarf is basically a star near death, singing its swan song. Massive stars, on the other hand, experience a more violent death in an enormous explosion called a supernova. The remnants of a supernova may become either a rapidly spinning neutron star or a black hole.

When astronomers in the 1960s turned radio telescopes on the sky for the first time, they discovered sources of radio waves. These were spread out along the Milky Way. When astronomers turned visible-light telescopes on those points in space, they found bright spots that might be distant stars. They named them quasi-stellar radio sources, or quasars for short. Quasars are extremely bright, billions of times brighter than our sun, and they are also extremely distant. It is believed that they are compact regions that draw their energy from supermassive black holes near the center of their galaxy. Their light takes billions of years to reach the earth, which makes them possible sources of information about the early stages of the universe.

Famous American astronomer, educator, and TV personality Carl Sagan once said, "A galaxy is composed of gas and dust and stars, billions upon billions of stars." Sagan's distinctive speech emphasized the letter "b" in "billions and billions," and the expression became a favorite catchphrase in astronomy discussions as well as a favorite target of comic performers. Sagan took all the comedy in good humor and gave the title *Billions and Billions* to his final book. Our title of this chapter is homage to the great astronomer and science popularizer Carl Sagan.

The truth is that some of the magnitudes in astronomy are just so difficult to grasp. Our Milky Way galaxy is said to contain four hundred billion stars. These are not just any celestial objects but real stars, each with its own solar system. That is a big number, four hundred billion stars; and yet our Milky Way is one of two hundred billion galaxies, a number that is likely to increase as we improve our ability to look deep into space. If we assume that the Milky Way is an average-sized galaxy, then we have eighty sextillion stars, or the number 8 followed by twenty-two zeros, or eighty billion trillion stars. No human mind can grasp these numbers or the cosmological vastness that spreads beyond our home planet. The cosmos is a wonderful world of fiction and, at the same time, a world of ultimate reality.

Much of the work in astronomy today centers around the Big Bang theory. Realistically, the concept of a beginning of the universe is troublesome. Does it make logical sense to say that it all started with a specific event? The first questions asked cannot be answered: What was there before? How is the law of conservation of energy fulfilled at time zero? Were the laws of physics created at that instant? How can our laws of physics guide us to an instant in time when they did not exist? There are too many logical contradictions and paradoxes.

However, if the Big Bang is the beginning of our part of the cosmos, the beginning of our visible universe, then, of course, the contradictions disappear. We can carry out a meaningful discussion, and we will say, in this spirit, that the Big Bang is an event that created our visible galaxies, our own Milky Way, and our solar system. The Big Bang has nothing to do with the beginning of the entire cosmos or the beginning of space-time or the beginning of all existence. We note that the *Encyclopedia Britannica* is careful enough to define the Big Bang as "a widely held theory of the evolution of the universe."[44] *Britannica* does not even mention the "beginning" of the universe; the Big Bang is just an evolutionary event.

NASA scientists propose that the Big Bang did not occur at a single point in space as an explosion but rather as a simultaneous appearance of space everywhere in the universe.[45] Before the Big Bang, space was no bigger than a point. In other words, space did not exist. The universe expanded from that zero-volume single point of origin. If we think of an inflating balloon, the radius of the balloon grows as the universe expands, but all points on the surface of the balloon (the universe) recede from each other. NASA scientists believe that the Big Bang model does not need to consider questions such as "What is the universe expanding to?" or "What caused the Big Bang?"

The idea of an expanding universe is a logical contradiction unless this is one of many universes. If "universe" means "everything there is," where is the universe expanding to? We can solve this contradiction by defining the cosmos as everything there is and stating that the cosmos consists of many universes. The theory of expanding universe would then apply only to our own visible universe.

---

44 http://www.britannica.com/EBchecked/topic/64893/big-bang-model.
45 http://map.gsfc.nasa.gov/universe/bb_concepts.html.

The Big Bang, according to NASA and other scientists, was an explosion that caused the creation of space-time from a singular point at time zero. So, time must have been created as a unidirectional entity, with a future but without a past. If it had a past, there could be no zero point, and no Big Bang. If space-time just before the Big Bang was just a singular point, didn't that point have to contain all the mass and energy that resulted in our vast universe? How can time be created at some instant? How can an event occur if time does not exist? The idea that everything, including space-time, was created at one instant from a singularity is probably a good solution to a fine mathematical equation, but it is not a very good idea of physical reality.

The most troublesome part of all this is that nothing is verifiable or falsifiable. You can detect some weak cosmic radiation somewhere in space and then say, "Oh yes, these are remnants of the Big Bang, this confirms the theory." Some cosmologists have gone as far as to suggest that the point of origin of the Big Bang was just a few millimeters across!

The problem here is that none of these ideas can be incorporated into our thinking in ways that the ideas can be experientially verified and validated in any way. Astronomy is a true science, like physics, but it seems that cosmology has made a full circle back to theology and metaphysics. In a world population of seven billion people, there is always going to be a large number of grateful readers willing to accept exciting ideas without much proof. Some of this is as exciting as the best science fiction—and as credible as science fiction.

The Big Bang is current orthodoxy in cosmology, but our skeptical voice is not a voice in the wilderness. Quite a few scientists around the world are beginning to question the basic premises of the theory. There are extreme views on both sides, just as there are moderate views. The idea that the Big Bang is not a creationist but an evolutionary event has great appeal, and we might predict that this idea will define the path of future research.

The Thinking Machine

Computers, Robots, and Artificial Intelligence

RISTOTLE'S SYLLOGISM IS the first known attempt to prescribe a logical process in a formalized, consistent method. In philosophy, it opened a great new path that led to epistemology, the systematic theory of knowledge. In science, it became the foundation of the scientific method.

Twenty centuries after Aristotle, Anglo-Irish writer and clergyman Jonathan Swift wrote his satirical novel *Gulliver's Travels* and described a machine that might "write books in philosophy, poetry, politics, laws, mathematics, and theology, without the least assistance from genius or study." Swift's unique mix of science fiction with a satirical critique of contemporary philosophy turned out to be more vision than fiction. Three centuries later, Swift's machine is owned and used by more than one billion people around the world, and society is structured around the computer in ways that were not foreseen in his time.

There are more fictional accounts of computers mimicking human intelligence. In Arthur Clarke and Stanley Cubrick's film, *2001: A Space Odyssey*, an intelligent spaceship computer named HAL maintains ship functions and interacts intelligently with the crew— and eventually attempts to take control of the spaceship. Interestingly, HAL's attempt to take control is not due to any consciousness but rather to malfunction.

The idea of creating thinking machines turned from fiction into scientific research in the mid-twentieth century. In 1951 British mathematician and computer pioneer Alan Turing proposed a test called "The Imitation Game", designed to test a machine's ability to exhibit humanlike intelligence. In our discussion of A. J. Ayer, we saw that Ayer had proposed that the distinction between a conscious man and an unconscious machine resolves itself into a distinction between different types of perceptible behavior, an argument that anticipated the Turing test of a machine's capability to demonstrate intelligence. In the Turing test, a panel of judges converses via keyboard with an unknown entity. If more than 50 percent of the panel votes that the entity is human when the entity is actually a computer, then the computer is said to have passed the Turing test.

In the 1970s a group of computer programs started to appear that employed small databases of English words combined with a series of rules for forming intelligent sentences. This was an attempt to cross the language barrier between man and machine. Many of these programs were written just for fun or as an attempt to pass the Turing test, and some became quite popular. The best way to test the capabilities of these programs is to use them, and the psychology department of the University of Toronto has compiled a list of links that provide free access to some of the most popular programs.[46]

The Turing test has fueled quite a bit of controversy, which is still going strong today, after sixty years or so. Opponents argue that the test is neither necessary nor sufficient for the demonstration of

46 http://psych.utoronto.ca/users/reingold/courses/ai/turing.html.

intelligence. The test encourages trickery, and many truly intelligent systems would fail the test. Supporters of the Turing test argue that the test is a valid scientific criterion.

The best-known challenge of the Turing test is John Searle's "Chinese Room". Searle is an American philosopher educated at the University of Wisconsin and Oxford. In 1980 Searle presented his Chinese room argument, intended to falsify the Turing test.

Assume that you do not speak Chinese, and you are in a room where someone slips a piece of paper with Chinese writing on it under the door. You notice there is a book on your table with "if-then" instructions on how to produce new Chinese symbols based on what you have received. You follow all these rules and produce a piece of paper that you slide under the door back to your friends outside. The same thing happens the next day. You receive the Chinese text, apply the rules to produce a response, and deliver your response in Chinese. You do not understand any of the words; you have just produced your text by applying the rules, and your text is in perfect Chinese and makes sense. The outside world is totally impressed with your understanding of Chinese, but you do not have a clue as to what you have written. You have only applied the rules as written in your instruction manual.

In all fairness to Turing, we might say that the Turing test is a conversational test of semantics in the early stages of artificial intelligence (AI), whereas Searle's Chinese Room is a challenge on the Turing test based on a stronger idea of AI, which would be applicable to later-stage AI developments. There is no reason why the mechanics of the Turing test could not be adapted to a stronger AI by enriching the man-machine dialogue. In other words, Searle's Chinese Room is more of a challenge on AI than a challenge on the Turing test. In any event, all these interesting debates set the stage for a vigorous expansion of artificial intelligence projects.

Back in 1985, twenty-two-year-old world chess champion Garry Kasparov played simultaneously against thirty-two computers in

Hamburg and won with a perfect 32–0 score, albeit with some difficulty. The four leading chess computer manufacturers had sent their top models, including eight models that were named "Kasparov." The difficult moment for the world champion occurred in a game against one of the Kasparov brand models. Kasparov thought that if he lost the game, people would think that he did so intentionally to get publicity for the manufacturer, so he doubled his efforts. He finally tricked his way into a win by offering a sacrifice that should have been refused. Kasparov says that "those were the good old days of man-machine chess."[47]

Eleven years later Kasparov was defeated by Deep Blue, a supercomputer made by IBM. This was the first time when a reigning world champion was defeated by a machine. IBM had built the machine specifically to beat Kasparov. The six-game series was finally won by the Russian grandmaster 3–1 with two draws. IBM redesigned Deep Blue, and the following year, the machine beat Kasparov 3–2 in a highly publicized game.

The legendary games between Deep Blue and Kasparov inspired artificial intelligence researchers, who saw a new, fertile ground for ideas and techniques. The psychological impact of having a computer beat the world champion was a tremendous boost for artificial intelligence. Chess provided a complex reasoning problem with specific performance criteria. Later advances in computer hardware made possible extensive search techniques and made computer chess less of a challenge for AI, but the benefits had been reaped.

The processing of natural language by computers has proven more difficult than anticipated. The main problem seems to be the assignment of meaning to language. Sometimes we say, "I can't find words to express my feelings." What this really means is that our complex psychology cannot be perfectly mapped into a symbolic system such as language. Human intelligence is a complex outcome of empirical

---

47 http://www.nybooks.com/articles/archives/2010/feb/11/the-chess-master-and-the-computer/.

sensations, not instantaneously programmed but experienced from infancy to adulthood through the five senses and shaped by our basic human needs and by the mind into perceptions that are evolving over time in unique ways. Because these perceptions are not analytic, they are impossible to describe in analytic formulations. This is the very reason machines, as we know them today, will not achieve human intelligence. Their humanlike intelligence may only approach the analytic part of our intelligence, even if we develop fuzzy languages, which are essentially analytic as well.

There are no ways to program speculation, intuition, and common sense, which are important parts of human thinking and decision making. Computers are good at chess because chess is analytic logic, but they are not as good at bridge and poker, where speculation and risk taking are important. Also, our common sense, by definition, has a social aspect, and computers are not exactly social animals. Instantaneous programming is no match for the slow, complex process of human learning. The boundary that will always separate man from machine is not processing capability but the machine's inability to experience direct sensation through the five senses. If a machine cannot feel the unique experience of a baby's first hug and kiss from her mother, if the machine cannot experience the adolescent's first feelings of erotic love, if pain and joy cannot be felt but only programmed, there will always be a difference; and the machine's intelligence will be analytic and nonempirical.

In spite of such objections, there is likely a big, bright future for AI, with great potential to benefit mankind. In 2014, Stanford University invited leading thinkers from several institutions to begin a hundred-year effort to study and anticipate how the effects of artificial intelligence will ripple through every aspect of how people work, live, and play. This effort, called the *One Hundred Year Study on Artificial Intelligence*, is the brainchild of computer scientist Eric Horvitz. Stanford scientists recognize that artificial intelligence is one of the most profound undertakings in science and one that will affect every aspect of human life. The studies will consider impacts of AI on automation, national security, psychology, ethics, law, privacy, democracy, and other issues.

In his book *Runaround*, written in 1942, famed science-fiction writer Isaac Asimov wrote his Three Laws of Robotics:

1. A robot may not injure a human being or, through inaction, allow a human being to come to harm.
2. A robot must obey the orders given it by human beings, except where such orders would conflict with the First Law.
3. A robot must protect its own existence as long as such protection does not conflict with the First or Second Law.

In his later fiction, Asimov added a Zeroth Law:

0   A robot may not harm humanity or, by inaction, allow humanity to come to harm.

Asimov's laws became quite popular in science fiction, as they were incorporated by other writers as well. It would not be such a bad idea if they were actually built into all AI. The big question, of course, is how do you model the meaning of harm and the judgment of harm's consequences?

Car manufacturer Honda began developing humanoid robots in the 1980s. In the year 2000, they presented their most advanced robot, ASIMO. Obviously named after Isaac Asimov, the name is also an acronym for Advanced Step in Innovative Mobility. The robot was designed as a multifunctional mobile assistant. ASIMO made many appearances around the world, entertaining and inspiring youngsters to study science and mathematics.

It is not unthinkable that an intelligent machine may be pro-grammed someday to write Mozart's forty-second symphony, with all the supreme imagination, drama, and divine harmonies of the Austrian genius. But what about Beethoven, whose symphonies show a strong evolutionary pattern? The third is very different from the first two, the fifth is different from all four before it, and the ninth is very different from all music ever written. Can such intellectual progression be recreated by a thinking machine? Beethoven's tenth symphony, written by BETHOVO the Robot, would be a great addi-tion to our listening repertoire!

Since BETHOVO is not here yet, we'd better stick with ASIMO. The droid cannot write music but can perform a whole lot of useful tasks. ASIMO can walk and climb stairs, can recognize voices and faces, and can understand preprogrammed postures and gestures. It has two camera eyes and can recognize the distance and direction of multiple moving objects. ASIMO has arms and hands and can turn on light switches, open doors, carry objects, and push carts. If you offer a handshake, ASIMO will recognize the gesture and will respond appropriately.

ASIMO is just four feet three inches tall, or 1.3 meters, and looks like a kid wearing a space suit. ASIMO's height is just right to look eye to eye with someone seated, as its main purpose was to help people confined in bed or a wheelchair. ASIMO can also do jobs that are too dangerous for humans, like going into hazardous areas, disarming bombs, or fighting fires. Now, this is the essence of robotics. It is a technology that, if developed properly, can be of enormous value to mankind.

Special-task robots are already helping mankind in big ways. Robotic arms are commonplace now in medical surgery rooms. They are manipulated by the doctor with a telemanipulator or via com-puter, and the surgeon does not even have to be in the room—he or she can be anywhere in the world. This is not science fiction; it is happening today! Mass-produced electronic circuits are now almost exclusively manufactured by robots. In car manufacturing, sequential robotic stations weld, glue, paint, and assemble automobiles. As the costs of robots decline, their applications in health, education, military, and manufacturing will be endless.

The need for more and more complexity in AI and robotics creates an enormous need for densely packed electronic circuitry. But the electronics industry is up to the task. Since the invention of the transistor back in the 1940s, transistor size and cost has declined exponentially, and manufacturers today are able to pack millions of transistors in an area the size of a fingernail. Gordon Moore, the American scientist and entrepreneur who founded Intel, predicted in 1965 that transistor density would double every two years or so. The exponential increase in transistor densities so far is consistent with Moore's Law. There is no doubt that the exponential progression predicted by Moore will hit a limit at some point, a limit imposed by the laws of physics. But new advances in nanotechnology and the prospect of a single electron transistor hold the promise of putting transistor densities on new tracks of exponential growth.

It can be predicted that logicians and philosophers will find an extremely fertile ground of ideas in AI that may reveal new aspects of how humans acquire knowledge and new ideas about the nature of meaning, truth, and reality. But we need to be careful to avoid the conclusion that a thinking machine can provide an absolutely objective view, a view independent of the human perspective. Any machine created by humans will necessarily carry the human perspective.

Scientific Method and Objectivity

A Logical Framework

N OUR DISCUSSION of Aristotle, we pointed out that he and others, such as Eratosthenes and Hipparchus, were the first scholars to establish a scientific method, the pursuit of knowledge through systematic observation, measurement, and experiment and the formulation, testing, and modification of hypotheses. Aristotle recognized that empirical input rather than pure reason is the only source of new knowledge. The basic steps of the scientific method have not changed much since Aristotle's time. There are typically four steps: (1) observation of a natural event, (2) formulation of a hypothesis to explain the event, (3) use of the hypothesis to predict the outcomes of new observations, and (4) performance of experiments to validate the predictions.

We can illustrate this with a simple example from everyday life: (1) We turn on a switch in the house and notice that the light does not come on. (2) A couple of thoughts come to mind. First, there may be a power outage. Second, the light bulb may be burned out. It seems easier to investigate the first thought, so we make the hypothesis that there is a power outage. (3) We make a prediction that if we try to switch on another light, the light will not come on, as per our hypothesis. (4) Finally, we perform the experiment by switching another light. The light fails to turn on, and our hypothesis appears to be correct. We can then try more switches to get a full validation of our hypothesis.

In real science, things are a bit more complicated. When a complex new scientific theory is published, peer review becomes an important aspect of the scientific method, and we might say that it is a crucial fifth step. A new scientific theory will be scrutinized quite thoroughly before gaining wide acceptance. Present-day scientific communities are so large and so open in their communications that peer review is extensive and comprehensive in scope and depth. In our day, peer review is dispersed among thousands of scientists worldwide who use every possible type of reasoning—inductive, deductive, and everything else—in order to falsify or validate a theory before final acceptance.

The evolution of science, especially in modern times, shows that the specific method and reasoning used by the scientist will be as multifaceted as possible, in anticipation of a very multidimensional and diverse peer review. It is no longer useful to argue about the advantages and disadvantages of inductive and deductive reasoning, because a diligent scientist today will use all methods and types of reasoning at his or her disposal.

Let us imagine that we are sitting around a table, discussing all the different scientific and philosophical ideas presented in this book, from Aristotle to Einstein to the present. We will endeavor to form a consensus around a commonsense concept of knowledge and scientific method. We are not driven by self-interest; we are free of ideological bias, and none of us represents any special-interest groups. We are driven only by our natural human curiosity to understand our world.

We shall base our discussion on the smallest number of assumptions possible, which are unprovable positions or postulates. In addition to supporting our discussion, these assumptions will become the foundation of an axiomatic logical reference frame, or simply "frame," on which our consensus will be developed.

The first unprovable idea that we must agree on is that there is a universe that exists independently of human existence. If humans become extinct, like dinosaurs did some sixty million years ago, the universe will continue to exist. Most sensible people should be able to agree that this idea is a good assumption—except, perhaps, for some philosophers! (But their healthy skepticism does not necessarily make them less sensible.)

Our first assumption makes it possible for us to define "universal reality." This is the reality that exists outside of the human mind, is independent of human observation and existence, and may be the same or drastically different from the reality perceived by the human mind. Universal reality is the Absolute, and God is the only being in the universe who is aware of universal reality. This assumption is part of a logical, not a theological or metaphysical, argument; that is, it does not predicate or imply the ontological existence of God. It may even be argued that God is ontologically identical with this universal reality, but there is no need to make this argument in this discussion.

Our second unprovable idea is that humans are unable to transcend their human perspective in their efforts to achieve awareness of universal reality. Any human knowledge of a reality will necessarily carry the human perspective, even if achieved by intelligent, thinking machines. By definition, that is not knowledge of universal reality. Humans and their machines can achieve only a human-perspective reality. We have, therefore, two types of reality: universal reality and human-perspective reality. These two types of reality cannot become identical, because humans are limited by the boundaries of the human perspective.

Our third postulate is that there is only one type of truth: the axiomatic logical-frame truth, a truth of a natural event achieved by inductive and deductive reasoning within a logical frame, which is based on a set of predefined axioms and precise definitions. The frame has logical coherence, and all ideas are provable from the axiomatic set. A proposition is valid if it is shown to be consistent with all axioms and all propositions that have been proven within the axiomatic frame. In that sense, any proposition can be shown to be valid within a suitably selected axiomatic reference frame. The validation or proof of an idea is derived from the application of analytic relations on axiomatic and definitional variables. Therefore, ideological disagreement cannot exist within a frame. Such disagreement is possible only across different frames and is due exclusively to different definitions and axiomatic sets. There are no other truths or variants of truths that exist outside of the frame, and there are certainly no truths that exist outside of the reasoning of the human mind. Consequently, the word "truth," as used in this discussion, will mean "axiomatic logical-frame truth." Similarly, the word "reality" will mean "human-perspective reality" unless otherwise noted.

The theory described has an epistemological character, but it also has a metaphysical underpinning: there is an ontological reality that is independent of the human mind. The "thing to be known" has characteristics that are independent of human existence and are therefore part of universal reality. Humans perceive certain characteristics of the object that are shaped by their sensory abilities and the processing abilities of the mind. These characteristics may be the same or different from the object's universal characteristics, and it is not possible for humans to know the difference. Any meaningful discussion of reality must be limited to human-perspective reality, which is the portion of universal reality that the human mind can know; and this portion may be qualitatively different from the ontological reality.

There is a difference between what one knows and what is to be known. Without this distinction there can be no science and no need for a discussion of scientific objectivity. The aim of science is to develop truths that will eventually become identical with reality, reaching the limits of human intellectual capabilities. Scientific work develops intermediate truths over time, toward the goal of achieving this identity of truth and reality. Any intermediate truths are believed to be approximations of reality.

These intermediate truths are similar to what Peirce believed to be consensus truths. In our discussion of Peirce, we referred to his view that there is only one reality. Truth and reality are determined by the consensus of those who investigate, and the force of consensus is beyond the power of each individual. Truth, therefore, is more of a social than an individual matter. The best representation of truth is the truth developed from a consensus formed out of the common elements of individual human perceptions. The peer review that we have described as the fifth step of the scientific method validates a theory, which then becomes a scientific truth. This view is close to Peirce's consensus idea.

Objectivity is at the heart of the scientific method. Science before Aristotle was quite often very much like philosophical and theological speculation, without much need of empirical proof. Since Aristotle, Eratosthenes, Hipparchus, and other Greek scholars of the time, we have developed a new approach to science wherein the truth of a scientific claim or hypothesis must be proven or must be provable. Objectivity, though not fully developed as a scientific concept, was implicit in all scientific work. It became an essential value in science, and it was what made science different from philosophical and theological speculation.

We must define scientific objectivity in a manner that makes logical sense and is practically useful. The dictionary definition of objectivity is "lack of bias or prejudice." The philosophical meaning of scientific objectivity is quite similar: objectivity expresses the idea that the claims, methods, and results of science are not and should not be influenced by particular perspectives, value commitments, community biases, or personal interests, to name a few relevant factors.

We will define objectivity as the central value of the scientific method that requires the absence of personal and apparatus bias and drives scientific truth toward a complete identity with human-perspective reality. In our discussion here, we will make a distinction between psychological and systemic objectivity. Psychological objectivity is the type of objectivity that has been described in dictionaries, dealing with bias and prejudice. Examples of psychological objectivity (or lack thereof) are plentiful: the persecution by the church of scientists and astronomers for holding views contradictory to religious beliefs; the long ideological battle between evolution and creationism; more recently, the ideologically charged debate of the findings of climate science; and situations of groupthink, such as the persistence of the scientific community on the existence of ether, followed by a quick acceptance of the flawed Michelson-Morley experiment. We have seen psychological bias even in Einstein, who compromised his own efforts toward a unified theory, because he could not accept uncertainty and quantum theory.

The debate on scientific objectivity has carried into the twenty-first century. In 2007 science historians Lorraine Daston and Peter Galison wrote an excellent five-hundred-page book simply titled *Objectivity*. Daston and Galison point out that during the Enlightenment, naturalists and scientific atlas makers practiced the truth-to-nature ideal, which involved active attempts to eliminate biases. In the nineteenth century, this scientific virtue developed into a more complete understanding of objectivity. The invention of photography allowed the abandonment of subjective and idealized representations of nature. But scientists realized in the twentieth century that simple mechanical depictions were inadequate, and they had to be combined with trained judgment to be of any scientific value.

There are many factors that affect scientific objectivity. Even the most diligent and bias-free scientists may be influenced by various factors that compromise their impartiality. Like all people, scientists carry their own personal perspectives. These may be ideological, psychological, or other types of attitudes shaped by social influences. Scientific objectivity requires freedom from personal, political, religious, and social preferences. Of the many psychological factors that influence objectivity, we will focus on groupthink, which has not been adequately recognized as a bias factor in the evolution of science.

Groupthink was defined by Irving Janis, an American research psychologist at Yale. Janis defined groupthink as "a mode of thinking that people engage in when they are deeply involved in a cohesive group, when the members' strivings for unanimity override their motivation to realistically appraise alternative courses of action."[48] In other words, the need for harmony and conformity within a group drives the selection of ideas, overlooking the critical appraisal of alternatives that may create conflict within the group. Interestingly, German nineteenth-century philosopher Nietzsche went so far as to say that madness is the exception in individuals but the rule in groups.

Think about the last time you were part of a group. Someone proposes an idea, and you may think that the idea needs some exploratory discussion about evidence, consequences, possible alternative ideas, and so on. However, you are reluctant to voice your opinion, seeing that most members agree with the idea and fearing that you might disrupt the harmony of the group. There may be other members in the group who feel the same way and refrain from voicing their objections. This gets even worse if there is a strong voice in the room, such as a chief executive officer, who makes his views known early in the session. The idea is quickly accepted without a critical appraisal of alternative ideas.

Janis's examples of group decisions that were influenced by groupthink were primarily focused on political decisions, such as Pearl Harbor, the Bay of Pigs invasion, and the Cuban missile crisis, as well as group decisions made in the corporate world. Other researchers, however, found instances of groupthink behavior in a much wider range of group settings than Janis had considered. American

---

48 Irving Janis, *Victims of Groupthink* (Boston: Houghton Mifflin Co., 1972), 9.

economist Daniel Klein and Swedish social researcher Charlotta Stern found evidence of groupthink in academic settings.[49]

In our review of the Michelson-Morley experiment, we found that scientific consensus was reached too quickly and was driven by the scientific community's anxiety with the uncomfortable concept of ether. The experiment failed to prove the existence of ether and left scientists without a transmitting medium for light. This was a lesser evil compared to the existence of ether and scientists embraced the result without asking too many questions about the basic premises and the integrity of the experiment. This attitude of the scientific community had many of the elements of groupthink behavior.

We saw that the geocentric idea of the universe prevailed for centuries with the support of theological dogmatism. Any alternative theories and dissenting voices were quickly suppressed by the church. The need for conformity was imposed, and a groupthink psychology developed in the scientific world.

Thousands of scientists around the world are currently engaged in a quest for a unified field theory, a general theory that includes all forces of the universe and all laws of physics in one theoretical framework. A theoretical standard model has been developed that has predicted many phenomena, has led to discoveries of new elementary particles, and yet is unable to account for gravity, the most familiar and intuitive force in nature.

Why must there be one law that explains everything? Are scientists underestimating the diversity of nature? Have they taken an uncontrolled flight into theory space? Are humans now able to see everything in the universe so they can make a theory of everything? Are we overlooking the alternative idea that nature is diverse and its phenomena have unique and maybe random behaviors? Can the universe be fit into a manmade theoretical straitjacket?

One wonders if there is a new groupthink situation going on in a large part of the scientific community. In our day, since the 1990s,

49 Daniel B. Klein and Charlotta Stern, *Groupthink in Academia* (The Independent Review, 2009).

the science of climate has been laden with controversy. The theory basically says that a gradual increase in global temperature has been observed and is due to greenhouse gases caused by human activities. The theory fell into the hands of political activist groups, and many scientists sided with these groups and engaged in ideological and political battles. There is nothing wrong with scientists being ideologically and politically active. Science does not live in an unworldly isolation; it is part of the real world. But when ideology compromises scientific objectivity, there is a problem.

In 2009 it was reported that scientists at the University of East Anglia's Climatic Research Unit in the United Kingdom confessed to throwing out raw temperature data that did not support the theory of global warming. Investigations revealed a major scientific scandal, probably the worst in world history. The scandal did much to damage the credibility of global warming, a crucial and legitimate scientific and social issue.

We can identify elements of groupthink here. In Janis's words: "Overestimate of the group's power and morality, including an unquestioned belief in the group's inherent morality, inclining the members to ignore the ethical or moral consequences of their action. Close-mindedness, including a refusal to consider alternative explanations and stereotyped negative views of those who aren't part of the group's consensus. The group takes on a win-lose fighting stance toward alternative views."[50]

There is no place for any bias in science, but ideologically motivated bias is one of the worst kinds of biases, just like religious bias was in Galileo's time. Ideology combined with groupthink is dynamite to the foundations of science!

We can distinguish another type of objectivity that we can call apparatus objectivity or, better yet, systemic objectivity. This type of objectivity relates to the ability of the observer's experimental apparatus

---

50 Irving Janis, *Groupthink: Psychological Studies of Policy Decisions and Fiascoes,* Second Edition, (New York: Houghton Mifflin, 1982), 247.

to observe the event and is the major factor that separates truth from reality, assuming that truth has been achieved in the absence of any personal bias. We must remember that scientific truth is not an ultimate truth but merely a truth derived within an axiomatic logical frame.

We have defined two types of reality. Which one are we going to use in our definition of objectivity? Can systemic objectivity be defined as the difference between truth and universal reality? Theoretically, yes, but practically, such a definition would be meaningless as this type of reality is beyond the boundaries of the human perspective and experience. We will therefore relate systemic objectivity with the striving for identity between truth and human-perspective reality.

The final goal and limit of this objectivity is the perfect identity of scientific truth with human-perspective reality. When truth becomes identical with human-perspective reality, perfect objectivity is achieved within the capabilities of the human intellect. This process may take several centuries for the most difficult problems of science, depending on the scale of the problem compared to the human scale. In this context, "scale" means the physical size of the event and the physical size of the measuring apparatus. For example, the problem of the classification of living species that Aristotle studied was neither microscopic nor macroscopic. It was a problem of the human scale. Direct observations could be made and recorded without the use of sophisticated apparatus. Also, with the invention of the telescope, Galileo expanded the observational scale of his apparatus and was able to make direct observations of celestial events.

Microscopic and macroscopic phenomena that occur on far smaller or far greater scales than the human scale are much more difficult. They will typically find speculative explanations until the experimental apparatus is improved by technology, a process that has taken several centuries in some sciences. We may therefore say heuristically that one of the factors affecting objectivity is the difference of scale between the measuring apparatus and the phenomenon investigated.

Accordingly, we may say that systemic objectivity is proportional to the ratio ρ_e/ρ_m, where ρ_e is the spatial size of the event and ρ_m is the resolution of the measuring apparatus. Similarly, in particle physics, we may say that systemic objectivity is proportional to the ratio λ_e/λ_m, where λ_e is the wavelength of the event and λ_m is the wavelength of the measuring apparatus. In microscopic phenomena, λ_m is typically larger than λ_e. As the size of λ_m is reduced and approaches the size of λ_e through technological innovation, the ratio λ_e/λ_m approaches unity, which means that we have achieved the maximum systemic objectivity as far as our observational ability is concerned. However, the fact that we have maximum or perfect objectivity does not guarantee that we have achieved scientific truth, as the event may still be unexplained in spite of the improved apparatus. Objectivity is therefore a necessary but not sufficient condition for scientific truth.

We often say that a scientific observer conducting an experiment must ensure that the observer or his or her apparatus does not in any way disturb the natural event under observation. The question is, can we meaningfully say that once all these nondisturbance measures have been taken successfully, our observer has an objective truth of the event? In relativity, the observer's view of the event will still be affected by the observer's position and motion. Once we define the observer's reference frame, we can describe the event in terms that apply to the frame. But the laws of physics must be the same in all frames. Philosophers and scientists have proven relativity in the observations of events but have not proven relativity in the laws of physics. They have assumed that the laws of physics are the same for everyone in every reference frame.

The postulate that all laws of physics are the same in all reference frames is not self-evident, is not intuitive, and is quite arbitrary. Humans have empirical knowledge from a very tiny part of the universe. We cannot know if the laws of our classical and modern physics will be confirmed by experiments performed at distant galaxies, if such experiments could ever be performed. We do not know if there is another type of radiation, not visible to us but perceptible to other beings, intelligent or not, that travels with speeds greater than the speed of light. Our laws of physics have been built on postulates of mathematics and physics that are valid in our own axiomatic reference frame and have been derived from extremely limited empirical input, filtered through our human perspective and a human scale of the universe.

We have argued that any scientific truth is an intermediate truth toward the goal of achieving perfect identity between truth and human-perspective reality within an axiomatic reference frame. We have also described the five steps of the scientific method, including the all-important step of peer review and acceptance by the scientific community, but we have said nothing about proof and the standards of scientific evidence.

Proof in the Euclidean sense does not exist in science. It only exists in mathematics and logic. The inductive process commits the logical fallacy of generalization, and yet it is one of the cornerstones of scientific discovery. There is really no problem with the use of inductive reasoning, unless we are trying to find the shortest straight line to the final truth. But there is no path—straight, curved, or broken— to the final truth, simply because the final truth does not exist, or it is very far away. Most scientific discoveries in history have been revised. Whatever final truth we can achieve, it will be a truth of the human perspective, and achieving it takes several inductive cycles over decades and centuries. The road from geocentrism to heliocentrism took seventeen centuries, and it is doubtful that we have a final truth yet. There are billions of trillions of species on our planet and in the universe, each having its own little or larger perspectives and its own final truths. Induction takes us no higher than to a likely truth based on our interpretation of empirical observations within the limitations of human perspective.

But there is another reason why a scientist does not care much about the deficiencies of the inductive or the deductive process. In addition to induction and deduction, a scientist will use his or her own intuition, speculation, common sense, and uncommon sense. These are logical processes used by scientists on a daily basis. The scientist is much more flexible, speculative, intuitive, and probabilistic than the philosophical formalism of the scientific method has portrayed to date. The role of intuition is especially important. The workings of intuition, imagination, and creativity can transcend those of pure logic, and we have seen that scientific discovery is often a triumph of intuition over logic.

Our common sense is not something fixed and absolute. It is like a living organism; it changes over time and is enriched with new empirical information. Quantum physics will be common sense to tomorrow's scientists. But our uncommon sense is also very important in the evolution of science. It is what we call "thinking out of the box." Uncommon sense is the ability to transcend the existing paradigm, to speculate and predict the new elements that will enrich our common sense in the future. The philosophy of science can only benefit from deriving empirical models that include common sense, intuition, and speculation in the process of scientific discovery.

INDEX